T0140410

Springer Theses

Recognizing Outstanding Ph.D. Research

Aims and Scope

The series "Springer Theses" brings together a selection of the very best Ph.D. theses from around the world and across the physical sciences. Nominated and endorsed by two recognized specialists, each published volume has been selected for its scientific excellence and the high impact of its contents for the pertinent field of research. For greater accessibility to non-specialists, the published versions include an extended introduction, as well as a foreword by the student's supervisor explaining the special relevance of the work for the field. As a whole, the series will provide a valuable resource both for newcomers to the research fields described, and for other scientists seeking detailed background information on special questions. Finally, it provides an accredited documentation of the valuable contributions made by today's younger generation of scientists.

Theses are accepted into the series by invited nomination only and must fulfill all of the following criteria

- They must be written in good English.
- The topic should fall within the confines of Chemistry, Physics, Earth Sciences, Engineering and related interdisciplinary fields such as Materials, Nanoscience, Chemical Engineering, Complex Systems and Biophysics.
- The work reported in the thesis must represent a significant scientific advance.
- If the thesis includes previously published material, permission to reproduce this must be gained from the respective copyright holder.
- They must have been examined and passed during the 12 months prior to nomination.
- Each thesis should include a foreword by the supervisor outlining the significance of its content.
- The theses should have a clearly defined structure including an introduction accessible to scientists not expert in that particular field.

More information about this series at http://www.springer.com/series/8790

Markus Rambach

Narrowband Single Photons for Light-Matter Interfaces

Doctoral Thesis accepted by
the University of Queensland, Brisbane, QLD, Australia

 Springer

Author
Dr. Markus Rambach
School of Mathematics and Physics
The University of Queensland
Brisbane, QLD, Australia

Supervisor
Prof. Andrew G. White
School of Mathematics and Physics
The University of Queensland
Brisbane, QLD, Australia

ISSN 2190-5053   ISSN 2190-5061   (electronic)
Springer Theses
ISBN 978-3-030-07311-4   ISBN 978-3-319-97154-4   (eBook)
https://doi.org/10.1007/978-3-319-97154-4

© Springer Nature Switzerland AG 2018
Softcover re-print of the Hardcover 1st edition 2018
This work is subject to copyright. All rights are reserved by the Publisher, whether the whole or part of the material is concerned, specifically the rights of translation, reprinting, reuse of illustrations, recitation, broadcasting, reproduction on microfilms or in any other physical way, and transmission or information storage and retrieval, electronic adaptation, computer software, or by similar or dissimilar methodology now known or hereafter developed.
The use of general descriptive names, registered names, trademarks, service marks, etc. in this publication does not imply, even in the absence of a specific statement, that such names are exempt from the relevant protective laws and regulations and therefore free for general use.
The publisher, the authors, and the editors are safe to assume that the advice and information in this book are believed to be true and accurate at the date of publication. Neither the publisher nor the authors or the editors give a warranty, express or implied, with respect to the material contained herein or for any errors or omissions that may have been made. The publisher remains neutral with regard to jurisdictional claims in published maps and institutional affiliations.

This Springer imprint is published by the registered company Springer Nature Switzerland AG
The registered company address is: Gewerbestrasse 11, 6330 Cham, Switzerland

Für meine Schwester und meine Eltern.

Supervisor's Foreword

Our world has been transformed by communication networks: in my grandparents' time, fast communication was expensive and rare, and it is now affordable and commonplace. Our current technologies can efficiently communicate classical information: if we could also communicate quantum information, then intriguing new capabilities are possible, from measurably secure communication, through unimpeachable voting, to networking quantum computers.

Such a quantum Internet will require quantum channels—typically single photons—and quantum memories. The last decade has seen great advances in developing quantum memories, with storage efficiencies and recall fidelities both exceeding 90%—at the cost of requiring very narrow linewidth photons, on the order of a few hundred kHz. However to date, photon sources have photon linewidths in the MHz to THz range; further, the "narrowish" sources typically have less than 50% duty cycles, due to their operation being divided into stabilisation and photon production stages. So, no quantum Internet!

In an experimental tour de force, Markus puts us back on the road, engineering an efficient light-matter interface. His photons have the narrowest linewidths ever produced, at 429 ± 10 kHz, are at the wavelength for the rubidium D_1 transition in quantum memories and are produced with a 100% duty cycle.

Markus' thesis beautifully demonstrates that new technology enables new physics. His source produces two different kinds of quantum frequency combs, shows up to 84 quantum revivals—well beyond the 9 previously observed—and demonstrates superb quantum interference. The most famous signature of quantum interference is the Hong-Ou-Mandel (HOM) effect, where photons arriving from separate directions at a beam splitter become glued together. Normally, this only takes place if the path difference between the photons is on the order of microns to millimetres—Markus sees it with more than a hundred metre difference between the photons, allowing quantum-secured, enhanced precision, distance sensing of sub-wavelength features.

But enough from me, turn the page and dive on in! You are in for a treat.

Brisbane, Australia   Prof. Andrew G. White
May 2018

Abstract

Quantum technologies are becoming the driving engine of physical innovations in the twenty-first century, leading science and humanity onto a path towards a technological revolution. Early experiments have shown the enormous potential of this field, but after more than two decades quantum technology is still without an outstanding candidate for its underlying architecture. The reason for this is specific inherent weaknesses, found in all quantum platforms available to date. For example, for photons, it is the difficulty to keep them stored locally and implement two-qubit gates due to their low interaction, while systems based on, e.g. atoms or ions, fall short on mobility and have a high experimental overhead, making them hard to transport. A promising path forward is hybridisation of quantum technologies, seeking to combine individual quantum architectures by transferring the information between two quantum systems of different types, harnessing their strengths, while circumnavigating their weaknesses.

We want to benefit from the high mobility and ease of transmission of photons for quantum communication and exploit the excellent read-out and storage capabilities of atomic qubits as a quantum memory. To realise this, efficient interaction between the two-qubit carriers is necessary.

The atomic transition usually has a much narrower bandwidth than single photons generated by spontaneous parametric down-conversion (SPDC), the current gold standard of producing high-purity heralded single photons at flexible wavelengths. A high-quality interface of light particles with atoms therefore demands matching the spectral properties between the photons and the resonances of the atomic species. As the manipulation of atomic transitions is limited, the solution is to significantly reduce the single-photon emission spectrum to fulfil the requirements. We achieve this by performing the down-conversion process inside an optical cavity in order to enhance the probability of creating the emitted pairs in the spectral and spatial resonator mode.

Previous cavity-based SPDC sources have achieved bandwidths comparable to atomic linewidths; however, this is not sufficient to be merged with high-efficiency storage schemes. The atomic memory with the highest demonstrated storage and recall fidelity is the gradient echo memory (GEM) based on rubidium. To achieve these outstanding fidelities, GEM requires photons at sub-natural linewidths, e.g. sub-MHz for rubidium. Additionally, the operation time of most sources is divided into stabilisation and photon production phases, resulting in typical duty cycles <50%. The so far narrowest photons from SPDC have bandwidths still well above a MHz, and only a few sources have demonstrated 100% duty cycle.

In this thesis, I realise an efficient light-matter interface to be used with a rubidium-based GEM. My source offers 100% duty cycle generation of sub-MHz single-photon pairs at the rubidium D_1 line using cavity-enhanced SPDC. I introduce a new technique—the "flip-trick"—using a half-wave plate inside the cavity to achieve triple resonance of the pump, signal and idler photons. This allows probabilistic creation of single photons at any given time (100% duty cycle) without compromising the achievable linewidth and enables the highest spectral brightness from a SPDC-based source to date. The double exponential decay of the temporal intensity cross-correlation function exhibits a bandwidth of 429 ± 10 kHz for the single photons, an order of magnitude below the natural linewidth of the target transition and well suited for the implementation of GEM. This is the narrowest bandwidth of single photons from SPDC reported so far.

The quantum nature of the source was confirmed by the idler-triggered second-order autocorrelation function at $\tau = 0$ to be $g_{s,s}^{(2)}(0) = 0.032 \pm 0.003$ for a heralding rate of 3.5 kHz, and antibunching below 0.5 was observed up to heralding rates of 70 kHz. The high multi-photon suppression of the source is matched by high indistinguishability of the photons, demonstrated in a Hong-Ou-Mandel (HOM) interference experiment with a visibility $V = 96.7 \pm 3.4\%$ of the central dip. In addition, the mode-locked two-photon state of the generated pairs leads to revivals of the HOM dip. We measured these revivals with up to 105-m path difference between signal and idler photons, where $V = 38.2 \pm 2.4\%$, giving independent proof of the exceptional coherence length of our photons.

The narrow bandwidth in combination with high brightness, multi-photon suppression and indistinguishability makes our system the perfect source for the future integration with GEM, one of the most promising schemes for quantum memories to date, or hollow-core glass fibres filled with rubidium gas to allow the construction of novel quantum logic gates. Furthermore, the extension of the photon wave packet over more than 100 m can easily cover a whole experimental set-up, making the photons an ideal candidate for measurements on quantum foundations, e.g. quantum causality.

Acknowledgements

First of all, I would like to apologise to all the people I might forget to thank here, be sure this does not belittle my gratitude towards you.

I would like to start off by thanking my advisory team: Andrew White, for his constant positive reinforcement of ideas concerning the experiments, his expertise and advice in many situations and for giving me the chance to work on a challenging project with many great people. And of course, his laughter, filling the corridors of Parnell with joy. Till Weinhold, for his patience in the laboratory and the daily discussions, without which the project would not have been a success, and for all the hospitality when I first arrived. Sandro, for playing the perfect counterpart to Andrew, and Marcelo, for helping out whenever needed.

To my parents: you made all this possible! With your unconditional support in every decision I made, no matter how far-fetched they were at times and even when I decided to go to the other side of the world to work towards my degree. The many many hours we spend on Skype and the stories that we shared this way. You were always there for me, in happy and sad times, and I will never be able to thank you enough for it.

I also want to thank my partner Anna from the bottom of my heart, for the encouragement and love she gave me during the last years. Coming home to you and the kids has been the highlight of every day and your compassion got me through many frustrating periods. I can always count on you to be next to me, through thick and thin. I cannot wait for our next adventure.

And finally, to everyone:
Thanks for the trouble!

Contents

1 **Introduction** 1
  1.1 A Hitchhiker's Guide to this Thesis............................ 4
  1.2 A Brief History of Narrowband Single Photon Sources
     from Spontaneous Parametric Down-Conversion 5
  References 11

2 **Theoretical and Experimental Foundations**...................... 15
  2.1 Optical Cavities 15
    2.1.1 Resonator Theory 16
    2.1.2 Mode-Matching...................................... 18
  2.2 Stabilisation Techniques 23
    2.2.1 Pound-Drever-Hall Frequency Stabilisation 24
    2.2.2 Reference Systems................................... 25
    2.2.3 Proportional-Integral-Derivative Controller 28
  2.3 Nonlinear Optical Processes 29
    2.3.1 Second Harmonic Generation 31
    2.3.2 Periodically Poled Nonlinear Crystals 35
    2.3.3 Spontaneous Parametric Down-Conversion 37
    2.3.4 Cavity-Enhanced Down-Conversion: Optical
        Parametric Oscillators 39
  2.4 Single Photon Source Metrics 42
    2.4.1 Spectral Brightness 43
    2.4.2 Intensity Cross-Correlation Function $G_{s,i}^{(2)}(\tau)$:
        Linewidth 44
    2.4.3 Intensity Auto-Correlation Function $g_{s,s}^{(2)}(\tau)$:
        Multi-photon Suppression............................. 47
    2.4.4 Hong-Ou-Mandel Interference: Indistinguishability 50
  References 56

3 Design of a Narrowband Single Photon Source 59
   3.1 Optical Parametric Oscillator Design 59
        3.1.1 Crystal 60
        3.1.2 Cavity Parameters 61
        3.1.3 Birefringence Compensation 65
   3.2 Photon Pair Filtering 70
        3.2.1 Mode-Cleaning Cavity 70
        3.2.2 Clustering Effect 74
   3.3 Electronics and Control 75
        3.3.1 Field-Programmable Gate Array 75
        3.3.2 Control Software 77
        3.3.3 Temperature Stabilisation 81
   3.4 Frequency Stabilisation 84
        3.4.1 Laser 85
        3.4.2 Cavities 87
        3.4.3 Atomic System 92
   3.5 Detectors 95
        3.5.1 Universal Photodetectors 96
        3.5.2 Single Photon Counting 97
   3.6 Experimental Setup 99
        3.6.1 Laser and Absolute Frequency Reference 99
        3.6.2 Conversion Setup 101
        3.6.3 Mode-Cleaning Setup 102
        3.6.4 Feedback Loops 102
   References 103

4 Single Photon Characterisation 107
   4.1 Classical Characterisation 107
        4.1.1 Intensity Cross-Correlation Function $G_{s,i}^{(2)}(\tau)$ 107
        4.1.2 Spectral Brightness 115
   4.2 Quantum Characterisation 116
        4.2.1 Multi-photon Suppression 116
        4.2.2 Indistinguishability 122
   References 128

5 Conclusions 131
   5.1 Summary 131
   5.2 Outlook 132
   References 134

Appendix A: HWP Characterisation 137

Appendix B: Photodetector Circuit Diagram 139

Curriculum Vitae 143

Chapter 1
Introduction

The history of mankind is a story of communication. Over centuries, the urge of human beings to exchange thoughts with each other over long distances has driven a variety of sociological and technological innovations: starting in ancient Persia and Egypt with letters written on the bark of trees or papyrus and transported by couriers, enriching cultural exchange between distant places, over the invention of the printing press for fast duplication of e.g. newspapers, or later the telephone for personal communication, up to today, where the whole world is just a switch away, either on radio, television, a computer or just a small mobile phone in your pocket.

One of the most challenging developments of the last decades is the rapid growth of local and global computer networks, most prominently the internet. Especially the increasing demand on sending and receiving data is pushing the connecting elements like modems (wired) or routers (wireless), to their boundaries. The fastest method for sending information to date is via optical glass fibre, where light pulses represent the data bits. This offers the advantages of high speed and low interaction of photons with the environment, resulting in fast communication with high noise suppression and small losses. Additionally, one cable can carry multiple wavelengths, dramatically increasing the possible transmission rate.

Quantum technologies will undoubtedly be involved in the future generations of networks, with many major companies and countries already investing in programs for their development. These quantum networks allow for transport of quantum information between different nodes, consisting of individual quantum systems [1–6]. They will include two main components, analogue to the classical case: a (quantum) computer with some type of storage device and a (quantum) communication mechanism, most likely single photons, designed to faithfully transmit (quantum) bits, called qubits. The interested reader can find further introductory information on the concept of qubits and quantum computation in general in various books, e.g. by Nielsen and Chuang [7] or Aaronson [8].

Photons are the natural choice for long distance information carriers in quantum communication as the qubit is easy to encode, e.g. in polarisation, the interaction with

© Springer Nature Switzerland AG 2018
M. Rambach, *Narrowband Single Photons for Light-Matter Interfaces*,
Springer Theses, https://doi.org/10.1007/978-3-319-97154-4_1

the environment is little, lowering losses, and the individual photon mobility is high. However, in order to communicate over large distances, unavoidable decoherence and fibre losses need to be compensated. This can be achieved by quantum repeaters, devices that faithfully restore the original information on a regular basis. The underlying concept of such a repeater is called entanglement swapping, a process where entanglement is transferred onto two particles from individual entangled subsystems (A/B) and (1/2). A joint measurement on A and 1 will project the two remaining particles B and 2 into an entangled state, hence the entanglement is "swapped". Quantum repeaters can be realised all-optical between pairs of photons [9, 10] or as a combination of memories and noiseless amplifiers between photons and other systems [1–3, 11–13].

As photons are generally difficult to store and keep in one location, an interface mapping the photonic qubit state onto a local memory, a node of the network, needs to be implemented. These nodes can be atoms [14–16], ions [17, 18], nitrogen-vacancy (NV) centres in diamond [19, 20], or rare-earth doped solids [21–23], to just name a few. However, since the spectral properties (especially the transition wavelength and its linewidth) of the storage systems are limited in their tunability, the single photons require precise quantum engineering to match the transition in the memory qubit, essential for efficient hybridisation of the two quantum platforms.

There are multiple ways of generating suitable single photons. The most prominent approaches are four-wave mixing (FWM) [24, 25], a non-linear process where photons of two wavelengths combine to produce two new wavelengths, trapped atoms [26] or ions [27, 28], where an excited particle decays into its ground state by emitting exactly one photon, nitrogen-vacancy (NV) centres in diamond [29], where defects in diamond are excited and can emit photons, and spontaneous parametric down-conversion (SPDC) [30–34], where one photon spontaneously decays into two single photons inside a non-linear crystal. While FWM and trapping particles require substantial experimental effort with low duty cycles and NV centres only achieved linewidths in the GHz regime so far, SPDC produces photons with the required spectral tunability and comparably low complexity of the experimental setup.

An ideal single photon source needs to fulfil certain characteristics. First, it needs to emit single photon states with a high probability while suppressing optical vacuum or multi-photon emission. This can be quantified in a Hanbury Brown and Twiss (HBT) setup [35]. Second, each photon should be the same as its predecessor or partner. This indistinguishability can be measured by the Hong-Ou-Mandel (HOM) effect [36]. Lastly, the source should be efficient, robust, easy to implement and on demand. Apart from the latter, SPDC offers all these properties with high multi-photon suppression, measurable by a near-ideal Glauber autocorrelation function [37, 38] around zero time delay $g_{s,s}^{(2)}(0) \leq 0.01$ [39], far below the classical limit of 1 [38, 40, 41], and exceptionally high indistinguishability, characterised by HOM interference visibilities $\geq 99.9\%$ [42]. Both these quantum features are important figures of merit for a reliable single photon source in quantum information and communication [43].

However, due to energy conservation combined with the spontaneous nature of the SPDC process and the properties of the nonlinear crystal, the frequency spectrum has a typical width of 100s of GHz up to THz, dependent on the specific crystal and the wavelengths involved. This is several orders of magnitude above the transition bandwidths in many atomic species used in memory schemes. Narrowing the spectrum is possible by spectral filtering [44], but this severely compromises the achievable brightness of the source. A different approach is to create the single photon pairs inside an optical cavity [30, 31, 33, 34, 45–56], mapping the spatial and especially the spectral properties of the resonator mode onto the photons and enhancing the output into the desired frequency mode(s).

One of the most promising architectures for quantum memories is the gradient echo memory (GEM) [15, 16, 57–59]. GEM is a coherent storage technique where a magnetic field gradient is applied to a cloud of atoms. The resulting position dependent Zeeman shift in the atomic energy levels allows to store different frequency components of the photons along the length of the ensemble. If the magnetic field is flipped at a certain time t, the atoms generate a photon echo at time $2t$. High recall fidelities up to 98% [16], coherence times up to 1 ms in rubidium [60] and exceptionally long coherent excitation times up to 6 h in rare-earth ion-doped crystals [23] have been demonstrated in this memory scheme. In order to achieve these high fidelities and storage times, the spectral width of the incoming photons not only needs to match the atomic transition, but be well below its natural linewidth, e.g. sub-MHz for a rubidium-based GEM which has a natural linewidth of 5.7 MHz [61].

The work presented in this thesis is primarily aiming at developing a quantum light source that is compatible with GEM using rubidium, enabling an efficient light-matter interface. All experiments with GEM to date are based on weak coherent pulses, demonstrating the potential of the scheme but, strictly speaking, not storing a single photon (Fock) state of light. For deeper understanding of the differences in states of light see e.g. [38, 62]. The reason for this experimental shortfall is the difficulty of building a bright, yet true single-photon source with the required spectral properties. This thesis fills this gap. We prove the single photon nature of the source in two independent experiments (heralded auto-correlation function and HOM interference) and characterise the classical properties, e.g. the photon bandwidth and spectral brightness.

We are using cavity-enhanced SPDC, also known as an optical parametric oscillator (OPO), described e.g. in [63], in a triply resonant configuration of pump, signal and idler wavelengths to generate photons resonant with the rubidium D_1 line at 795 nm and surpass the bandwidth of the transition by more than an order of magnitude, i.e. ideal conditions for high storage and recall probabilities in the memory. The optical cavity design determines the output spectrum by enhancing the emission of the SPDC into cavity modes, while suppressing unsupported frequencies. In order to select one specific spectral mode resonant with the atomic species, the output requires an additional filtering step. Nevertheless, the obtained brightness is much higher than brute-force filtering of free-space SPDC or other methods. The triple resonance condition of the OPO allows for 100% duty cycle, meaning the source can probabilistically (spontaneously) produce single photon pairs at any given time.

Although the source is designed for integration with GEM, the photons can also be utilised in other applications like quantum information processing and quantum foundations. The source may be combined with gas-filled hollow-core photonic crystal fibres [64, 65], in order to achieve high cross-phase modulations [66, 67] or precise spectroscopy [68–71]. Especially the cross-phase modulation, where the presence of one photon can affect the phase of another via the optical Kerr effect, is of high interest in the field of optical quantum computing. This nonlinear effect forms the basis for a two-photon gate, as engineering deterministic interactions between photons is difficult due to their weak interaction and needs to be mediated by a nonlinear medium [72]. However, no system has yet shown a Kerr effect of sufficient strength. The exceptional temporal and spatial length of the single photon wave packet can furthermore be useful in quantum foundation experiments. In order to perform quantum computations without definite causal structure, also known as the quantum switch [73, 74], it is essential for the photon to extend over the whole experimental setup to erase the path information that could be acquired otherwise. Thus, the qubit becomes entangled with the circuit structure, revealing unexplored aspects of quantum theory.

1.1   A Hitchhiker's Guide to this Thesis

The thesis is organised in the following way:

This chapter provides an introduction to the thesis, starting with a motivation on the presented work. Then, a brief literature review identifies important parameters to explain where our source improves the field. More than 20 articles from the past 15 years are included to classify our work as complete as possible.

Chapter 2 explains important theoretical and experimental concepts, necessary to derive and understand the results obtained in this work. First, the theory of optical resonators, including their spectral and spatial properties, is presented. Afterwards, stabilisation techniques utilised in the experiment are discussed, focussing on the Pound-Drever-Hall technique and the Proportional-Integral-Differential control algorithm for frequency stabilisation. Next, nonlinear optical processes are introduced, with detailed descriptions of second harmonic generation and spontaneous parametric down-conversion followed by the combination of optical cavities and SPDC: optical parametric oscillators. The chapter ends with the derivation of single photon metrics, namely linewidth ($G_{s,i}^{(2)}(\tau)$), spectral brightness, multi-photon suppression ($1 - g_{s,s}^{(2)}(0)$) and indistinguishability (HOM interference).

Chapter 3 describes the design considerations and preliminary characterisations necessary to construct a narrowband single photon source. The first two sections summarise the major optical elements in the experiment, most prominently the OPO with a novel technique of birefringence compensation, its parameters and possible filtering techniques towards a single-mode output spectrum. This is followed by the electronic control systems in place in order to implement the various stabilisation loops. Here, detailed characterisation of the individual components like cavities or temperature controllers is paired with a description on how to operate the software for optimal frequency control. After a short treatise on utilised photon detectors, the chapter concludes with a summary of the complete experimental setup and feedback loops, illustrating how all components fit into place and how they work together.

Chapter 4 presents the results obtained for the characterisation of the photon pair source. It is divided in two major sections: part one describes classical characterisations like linewidth and spectral brightness, and the methods to obtain their values from the measured data in post-processing. The source is shown to be exceptionally bright and narrow in frequency. In part two, we explore the quantum nature of the source, with a deep analysis on the dependence of the auto-correlation function on various parameters. We further investigate the indistinguishability of the single photons in the case of a multi-mode frequency spectrum, demonstrating HOM dip revivals with high visibility and independently proving the long coherence time.

Chapter 5 summarises the results, discusses possible implications and gives a brief outlook on current work in progress and upcoming projects.

1.2 A Brief History of Narrowband Single Photon Sources from Spontaneous Parametric Down-Conversion

Over the last 15 years, multiple research groups all around the globe achieved narrowband emission of single photon sources. Spontaneous parametric down-conversion (SPDC) inside an optical cavity, also known as an optical parametric oscillator (OPO), is a good candidate for this purpose: SPDC is the current gold standard of producing high-purity heralded single photons at flexible wavelengths and the cavity enhances emission into certain spectral modes, tailoring the photon into the desired shape. In their groundbreaking work at the end of the last millennium, Ou and Lu expand the preliminary theoretical framework of SPDC inside an optical cavity from squeezing [75] to narrowband emission and demonstrated their predictions in an experiment [30, 45]. The authors achieved single-mode operation from a semi-monolithic standing wave cavity combined with a mode-cleaning cavity (MCC). The frequency mode had a width of $\Delta\nu_{SP} = 44$ MHz, narrowing the emission from single-pass operation by more than four orders of magnitude.

In recent years, most efforts to advance the field came out of China [48, 49, 52, 76, 77], Germany [31, 54, 56, 78–81] and Spain [33, 82–85], producing photon sources at various wavelengths resonant with atomic transitions in different memory schemes. During this time, new techniques for birefringence compensation and filtering were developed, resulting in higher spectral brightness, and passing the threshold to make OPO-based photon generation attractive for quantum information processing and quantum communication [1, 4, 6].

The next couple of paragraphs will analyse existing single photon sources based on SPDC (up to ∼100 MHz linewidth) for their similarities and differences and present where the work described in this thesis fits into the field. The review is divided in two big topics: first we will discuss design considerations of the utilised cavities, including the emission wavelength and linewidth, type, compensation method and duty cycle (closely related to the locking scheme). The second part will examine important characteristics of the sources, looking at the spectral brightness, auto-correlation ($g_{s,s}^{(2)}(0)$) and non-classical Hong-Ou-Mandel (HOM) interference, and also mention special features of some OPOs.

The first property to think about when building a narrowband OPO is its purpose. Most commonly, these sources are designed to interact with some type of atomic species, often Alkali atoms, especially rubidium and caesium [31, 32, 34, 46, 48, 49, 56, 76, 77, 81, 83, 84], or rare-earth ion-doped crystals [33, 54, 85], forming the basis of a quantum memory. Naturally, the corresponding requirements on the linewidth vary significantly dependent on the species and the memory scheme the photons are used in. This means that for storage applications, the single photon linewidth only needs to fulfil the demands of the memory and does not necessarily require to be as small as possible, although narrower photons usually offer higher storage fidelities. However, for different applications, e.g. experiments in quantum foundations, higher delocalisation of the photon wave packet and therefore long coherence times are always favourable.

One of the most promising candidates for quantum memories today is based on the photon-echo effect, demonstrated in various atomic species [5, 15, 16, 21–23, 44, 57, 86, 87]. The required linewidths vary from ∼100 MHz for neodymium-based crystals [44] down to sub-natural linewidth levels <1 MHz for rubidium [15, 16]. Generally, narrow linewidths below the memory bandwidth allow longer storage times and higher storage and recall fidelities and are therefore beneficial for efficient light-matter interfaces. Prior to the work presented in this thesis, Fekete et al. [33] demonstrated the narrowest photons from SPDC with $\Delta\nu_{SP} = 1.6$ MHz with a duty cycle of 55%, used for the integration with a praseodymium-doped crystal memory [85]. Our photon source is designed to work with a subset of the photon-echo technique, so-called gradient echo memories (GEM), in rubidium, with reported recall fidelities of up to 98% [16] and storage times up to 1 ms [60]. In GEM, an ensemble of atoms is frequency shifted to create a gradient in the transition frequencies along the length of the ensemble. Flipping the gradient after a certain amount of time t is generating the photon echo after $2t$. This scheme requires a spectral bandwidth of the photons well below the natural linewidth, e.g. sub-MHz for the rubidium D_1 transition with $\Gamma = 5.8$ MHz [61], in order to achieved the high fidelities and storage times. Hence, our source, with a demonstrated photon bandwidth of $\Delta\nu_{SP} = 429$ kHz at 795 nm, is perfectly suited for integration with GEM.

The second important consideration before starting to build the source is the design of the cavity, including the compensation method, duty cycle and, closely related, the resonance condition. The two main approaches are bi-directional standing-wave cavities (SW) [31, 46, 48, 52, 53, 56, 76, 77, 88], where the nonlinear crystal is traversed twice per round-trip, and travelling-wave resonators (TW) [33, 34, 47, 83–85], most importantly bow-tie cavities, single directional with only one pass of the light through the crystal per round-trip. TW cavities offer some advantages like lower losses per round-trip, increasing the finesse, or easy spatial distinction between incoming and outgoing beams, beneficial for detection of error signals and lowering the observed rate of unwanted photons created by back reflections of the pump light. On the other hand, their realisation can be more challenging as SW cavities consist of as little as two mirrors and a nonlinear crystal, sometimes implemented as coatings of one [48] or both sides of the crystal (monolithic cavities) [54, 88].

Apart from a few exceptions [30, 33, 49, 53, 85], most of the OPOs in the literature use type II down-conversion, where the created signal and idler photon are orthogonally polarised. This has the benefit of easy and deterministic separation of the pair on a polarisation beam splitter behind the cavity, however, achieving double or triple resonance is challenging due to the birefringence introduced by the crystal. In order to overcome this issue, different solutions have been realised: compensation crystals [31, 46, 52, 84], the clustering effect [33, 48, 54, 56, 77, 85, 88] or in our case, a novel technique utilising a half-wave plate [34]. Compensation crystals are a straight forward approach that cancels out the introduced birefringence through an additional crystal with its optical axis rotated by 90°. However, the extra losses introduced by the crystal and its surfaces reduce the finesse and broaden the linewidth. The clustering effect essentially anti-compensates the birefringence, accomplishing a high mismatch between the free spectral range (FSR) of signal and idler. Precise tuning of the crystal angle and the temperature creates small resonant clusters, each consisting of a few frequency modes, and results in a pre-filtered output spectrum. Nevertheless, meaningful utilisation of the clustering effect requires either vastly different frequencies of the photons [33, 85], small cavities of length comparable to the crystal length [53, 56, 88] or a combination of the two [54]. The narrowest linewidth achieved for degenerate photons from SPDC so far is $\Delta \nu_{SP} = 8.3$ MHz [88], still a factor of 50 higher than our source and therefore unsuitable for GEM.

The presented results in this thesis are based on an OPO using a HWP at 45° to compensate the intra-cavity birefringence. The HWP rotates the polarisation of the single photons by 90° every physical round-trip. This effectively doubles the length of the cavity and cancels out birefringence while leaving the linewidth unchanged compared to the OPO without compensating elements. Additionally, it allows easy implementation of triple resonant operation of the cavity, a property where many sources fall short. In this regime, the experiment can probabilistically produce and detect single photon pairs at any given time, further referred to as 100% duty cycle. Unfortunately, this important feature is almost never addressed in the literature for duty cycles below maximum, with only two references reporting on their actual values [33, 52] significantly below 100%. We can only infer shortcomings from statements on the resonance condition, as an OPO that is not resonant with the pump frequency always requires a locking cycle, a period during which no photons can be detected due to stabilisation at the single photon frequency. The triple resonance of pump, signal and idler photons has been demonstrated just a few times so far [31, 56, 81], with only Scholz and coworkers [31] achieving a linewidth comparable to our source, but still nearly an order of magnitude higher.

Table 1.1 Comparison of cavity design parameters in the literature, sorted by achieved linewidth. Note, narrowest linewidth is not necessarily the goal of every reviewed article. λ, wavelength; $\Delta \nu_{SP}$, single photon linewidth; DC, duty cycle; SM, single mode output; CM, birefringence compensation method; BT, bow-tie cavity; SW, standing wave cavity; M, monolithic cavity; WGMR, whispering-gallery mode resonator; n, no; y, yes; HWP, half-wave plate compensation method; CC, compensation crystal; ClE, clustering effect

Source	$\Delta \nu_{SP}$ [MHz]	DC [%]	λ [nm]	SM	Cavity type	CM	Year
This work	0.43	100	795	n	BT	HWP	2017
Fekete et al. [33], Rieländer et al. [85]	1.6/2.9	55	606/1436	n (few)	BT	ClE	2013
Scholz et al. [31, 78, 79]	2.7	100	894	y	SW	CC	2009
Schunk et al. [80, 81]	>6.6	100	795/1610 or 895/1312	y	WGMR	ClE	2016
Wolfgramm et al. [82, 84]	7	–	795	y	BT	CC	2008
Neergaard-Nielsen et al. [47]	8	–	860	n (few)	BT	–	2007
Zhou et al. [52]	8	2.5	1560	n	SW	CC	2014
Chuu et al. [88]	8.3	–	~1064	y	M	ClE	2012
Bao et al. [48], Yang [76]	9.6	–	780	y	SW semi-bulk	ClE	2008
Tian et al. [77]	15	–	795	y	SW	ClE	2016
Wang et al. [49]	21	–	780	y	SW	–	2010
Kuklewicz et al. [46]	22	–	795	n	SW	CC	2006
Haase et al. [83]	22	–	850/854	y	BT	–	2009
Ou and Lu [30, 45]	44	–	855	y	SW	–	1999
Luo et al. [54]	66	–	890/1320	y	M	ClE	2015
Ahlrichs and Benson [56]	100	100	894	y	SW	ClE	2016
Monteiro et al. [53]	116	–	1547/1569	y	SW	ClE	2014

A summary of the reviewed papers and their spectrum and cavity design parameters is presented in Table 1.1. The values for the duty cycle are given where available, all other numbers indicated by "–" are definitely <100% but could not be determined from the literature. Most of the sources achieved single-mode output through post-filtering of some kind, described in further detail later in the section. We have not demonstrated this so far, but all necessary components are built and characterised and we are currently working on including them in the experimental setup.

After finding similarities and differences in the designs and purposes of existing sources, the rest of this section aims at analysing the literature for characteristic features of the OPOs, namely spectral brightness, auto-correlation around zero time delay, HOM interference visibility and special details, making the sources unique in some way.

The spectral brightness (SB) is an important figure of merit of a narrowband single photon source, as it allows for meaningful comparison of sources from different architectures and filtering methods and describes if the enhancement of the emission by an OPO outperforms the brute-force approach of passive filtering of very bright free-running sources. However, it can be difficult to compare the SB presented by individual articles in a meaningful manner, as it is calculated in various different ways. Sometimes the SB is used as a synonym of the spectral generation rate inside the cavity, which is orders of magnitude higher. In other reports, losses from e.g. coupling, optics and filters are taken into account, again, leading to values many orders of magnitude above the actually detected rate. Throughout this thesis, we define the spectral brightness as the coincidence rate after the first optical fibre, per MHz bandwidth and per mW pump power, as this describes the actual SB of two-photon events that can be used in further experiments (analogous to [89]). Hence, we only account for the detection efficiency when calculating the values and solely compare literature that allows for an estimation per our definition. We believe this is the most honest interpretation and allows comparison to sources from other architectures. Over the last decade, the brightness of OPOs has been improved from the first reported $0.07 \frac{\text{photon pairs/s}}{\text{mW} * \text{MHz}}$ by Kuklewicz et al. [46], over the former best of $1410 \frac{\text{photon pairs/s}}{\text{mW} * \text{MHz}}$ by Chuu and coworkers [88], to the value in this thesis of $3900 \frac{\text{photon pairs/s}}{\text{mW} * \text{MHz}}$. For comparison, the SB of the currently brightest free running sources from quantum dots is still well below our value, e.g. $15 \frac{\text{photon pairs/s}}{\text{mW} * \text{MHz}}$ inferred from the work by Loredo et al. [89]. The exceptionally high spectral brightness of our source enables high repetition rates of experiments on light-matter interfaces with atomic memories like GEM, limited only by the preparation time of the memory (~ 100 ms [60]).

In terms of quantum characteristics, the multi-photon suppression and indistinguishability of the produced photons are the most significant features. The multi-photon suppression of a source describes its ability to create a single photon Fock state $|n = 1\rangle$ without higher order contributions. It can be quantified by the auto-correlation function [37, 38] around zero time delay $g_{s,s}^{(2)}(0)$ in an Hanbury Brown and Twiss (HBT) setup [35]. In the experiment, the signal photon is separated on a 50/50 beam splitter and both output ports are detected individually. A simultaneous coincidence indicates an unwanted multi-photon event from the OPO, reflected by an increasing value for $g_{s,s}^{(2)}(0)$. Obtaining this quantity in the field of narrowband single photons for meaningful comparison is non-trivial, as the auto-correlation function is defined at equal arrival times, but in an actual experiment, a coincidence time window has to be implemented, significantly influencing the achievable value. Unfortunately, it is not possible to "normalise" the numbers in the literature to a uniform time window, but more important than the actual value is showing $g_{s,s}^{(2)}(0) \ll 0.5$, indicating dominance of the single-photon Fock state [38]. There is only a small subset of articles [31, 34, 54, 84, 85] reporting on their autocorrelation measurement, but all achieve values in the low percent range: $0.012 \leq g_{s,s}^{(2)}(0) \leq 0.040$. Our OPO fits well into the literature with $g_{s,s}^{(2)}(0) = 0.032 \pm 0.003$, 277 standard deviations below the classical limit of $g_{s,s}^{(2)}(0) = 1$ [38].

The indistinguishability of the photons, important for e.g. optical quantum computation and communication, can be determined in a HOM interference measurement [36] and is usually quantified by the visibility V. It is, within error margins, independent of the chosen coincidence window for small photon detection rates, hence, comparing visibilities is a straight forward task. Again, all reported measurements [31, 48, 82, 85] show high visibilities of the central interference dip between 90% [82] and 97%, demonstrated in this thesis. This indicates large indistinguishability as expected by SPDC-based sources, ideal for entanglement swapping schemes in quantum repeaters or memories. An additional feature of our OPO in multi-mode operation is the revival of HOM interference dips with a period of $\frac{1}{2\nu_{FSR}} = 4.14$ ns, with ν_{FSR} the free spectral range of the cavity. This arises from the unique spectral and temporal shape of the photons, given as a superposition of individual narrow-band modes within the phase-matching envelope of the nonlinear crystal [90, 91]. We observe slowly decaying HOM interference visibilities as the delay increases, with $V = 38\%$ for signal and idler photon arriving at the beam splitter 105 m apart

Table 1.2 Comparison of the single photon source characteristics in the literature. SB, spectral brightness; $g_{s,s}^{(2)}(0)$, auto-correlation value at zero time delay; HOM, Hong-Ou-Mandel interference visibility; n/a, no available data; HWP, half-wave plate; MCC, mode-cleaning cavity; WGMR, whispering-gallery mode resonator

Source	SB $\frac{\text{photon pairs/s}}{\text{mW} * \text{MHz}}$	$g_{s,s}^{(2)}(0)$	HOM [%]	Special feature
This work	3900	0.032	97	HWP, MCC, HOM dip revivals
Fekete et al. [33], Rieländer et al. [85]	11	0.035	n/a	Type I
Scholz et al. [31, 78, 79]	330	0.012	n/a	MCC
Schunk et al. [80, 81]	–	n/a	n/a	Large tunability, WGMR, Type I
Wolfgramm et al. [82, 84]	28	0.040	90	Rubidium filter
Neergaard-Nielsen et al. [47]	200	n/a	n/a	Filtering line
Zhou et al. [52]	0.11	n/a	n/a	MCC
Chuu et al. [88]	1410	n/a	n/a	
Bao et al. [48], Yang [76]	6	n/a	97	Etalons
Tian et al. [77]	–	n/a	95	Filtering line
Wang et al. [49]	–	n/a	n/a	MCC, Type I
Kuklewicz et al. [46]	0.07	n/a	n/a	Notlocked
Haase et al. [83]	1.0	n/a	n/a	Filtering line
Ou and Lu [30, 45]	–	n/a	n/a	Type I
Luo et al. [54]	–	0.020	n/a	Not locked, waveguide
Ahlrichs and Benson [56]	33.7	n/a	96	Filtering line
Monteiro et al. [53]	–	n/a	n/a	Type 0

from each other, far beyond everything demonstrated in the literature and illustrating the exceptionally long coherence length of the created pairs.

Single mode operation, briefly discussed earlier in this section, is another crucial feature of the OPOs in order to efficiently interact with the atomic transition. Usually, this is achieved with one or multiple additional cavities for mode-cleaning or etalons in a filtering line [31, 47–49, 52, 56, 77, 83]. Careful design of these cavities allows high transmission of the desired mode >90% with reasonable extinction ratios, but often more than one system for filtering is required, adding complexity to the setup. Wolfgramm and coworkers showed that it is also possible to use the atomic species as a filter, with high extinction ratio ≥35 dB, but poor transmission ≤10% of the desired mode [84]. We have not yet implemented a filtering step, but preliminary characterisations and calculations show (near) single mode operation of the source with only one triangular filtering cavity and a fibre-to-fibre transmission of 45%.

The analysed OPO characteristics are shown in Table 1.2. An entry of "−" for the spectral brightness means that either the article did not supply enough information to re-calculate the value or it was not specified at all. An "n/a" in the $g_{s,s}^{(2)}(0)$ or HOM column refers to no available data found in the literature.

Although using a slightly different and far more challenging design of a monolithic whispering-gallery mode resonator (WGMR) made of the nonlinear crystal, the recent work by Schunk and coworkers [80, 81] is included in the review due to its high potential in the field. Here, the authors demonstrate 100% duty cycle, linewidths down to 6.6 MHz, fundamentally limited by losses inside the resonator material, and a wavelength tunability of the photons over 100s of nanometers at a constant pump wavelength (532 nm) achieved by temperature tuning of the WGMR. However, multi-photon suppression and indistinguishability experiments have yet to be published.

References

1. Duan, L.M., Lukin, M.D., Cirac, J.I., Zoller, P.: Long-distance quantum communication with atomic ensembles and linear optics. Nature **414**, 413–418 (2001)
2. Simon, C., et al.: Quantum repeaters with photon pair sources and multimode memories. Phys. Rev. Lett. **98**, 190503 (2007)
3. Chou, C.-W., et al.: Functional quantum nodes for entanglement distribution over scalable quantum networks. Science **316**, 1316–1320 (2007)
4. Kimble, H.J.: The quantum internet. Nature **453**, 1023–1030 (2008)
5. Tittel, W., et al.: Photon-echo quantum memory in solid state systems. Laser Photonics Rev. **4**, 244–267 (2010)
6. Sangouard, N., Simon, C., de Riedmatten, H., Gisin, N.: Quantum repeaters based on atomic ensembles and linear optics. Rev. Mod. Phys. **83**, 33–80 (2011)
7. Nielsen, M.A., Chuang, I.L.: Quantum Computation and Quantum Information, 10th edn. Cambridge University Press, New York (2011)
8. Aaronson, S.: Quantum Computing since Democritus. Cambridge University Press (2013)
9. Munro, W.J., Harrison, K.A., Stephens, A.M., Devitt, S.J., Nemoto, K.: From quantum multiplexing to high-performance quantum networking. Nat. Photonics **4**, 792–796 (2010)

10. Munro, W.J., Stephens, A.M., Devitt, S.J., Harrison, K.A., Nemoto, K.: Quantum communi-cation without the necessity of quantum memories. Nat. Photonics **6**, 777–781 (2012)
11. Briegel, H.-J., Dür, W., Cirac, J.I., Zoller, P.: Quantum repeaters: the role of imperfect local operations in quantum communication. Phys. Rev. Lett. **81**, 5932–5935 (1998)
12. Ho, J., Boston, A., Palsson, M., Pryde, G.: Experimental noiseless linear amplification using weak measurements. New J. Phys. **18**, 093026 (2016)
13. McMahon, N.A., Lund, A.P., Ralph, T.C.: Optimal architecture for a nondeterministic noiseless linear amplifier. Phys. Rev. A **89**, 023846 (2014)
14. Phillips, D.F., Fleischhauer, A., Mair, A., Walsworth, R.L., Lukin, M.D.: Storage of light in atomic vapor. Phys. Rev. Lett. **86**, 783–786 (2001)
15. Hosseini, M., Sparkes, B.M., Campbell, G., Lam, P.K., Buchler, B.C.: High efficiency coherent optical memory with warm rubidium vapour. Nat. Commun. **2**, 174 (2011)
16. Hosseini, M., Campbell, G., Sparkes, B.M., Lam, P.K., Buchler, B.C.: Unconditional room-temperature quantum memory. Nat. Phys. **7**, 794–798 (2011)
17. Kielpinski, D., et al.: A decoherence-free quantum memory using trapped ions. Science **291**, 1013–1015 (2001)
18. Schindler, P., et al.: A quantum information processor with trapped ions. New J. Phys. **15**, 123012 (2013)
19. Fuchs, G.D., Burkard, G., Klimov, P.V., Awschalom, D.D.: A quantum memory intrinsic to single nitrogen-vacancy centres in diamond. Nat. Phys. **7**, 789–793 (2011)
20. Heshami, K., et al.: Raman quantum memory based on an ensemble of nitrogen-vacancy centers coupled to a microcavity. Phys. Rev. A **89**, 040301 (2014)
21. Afzelius, M., Simon, C., de Riedmatten, H., Gisin, N.: Multimode quantum memory based on atomic frequency combs. Phys. Rev. A **79**, 052329 (2009)
22. Rieländer, D., et al.: Quantum storage of heralded single photons in a praseodymium-doped crystal. Phys. Rev. Lett. **112**, 040504 (2014)
23. Zhong, M., et al.: Optically addressable nuclear spins in a solid with a six-hour coherence time. Nature **517**, 177–180 (2015)
24. Zhao, L., et al.: Photon pairs with coherence time exceeding 1 μs. Optica **1**, 84–88 (2014)
25. Liao, K., et al.: Subnatural-linewidth polarization-entangled photon pairs with controllable temporal length. Phys. Rev. Lett. **112**, 243602 (2014)
26. Rosenfeld, W., et al.: Towards high-fidelity interference of photons emitted by two remotely trapped Rb-87 atoms. Opt. Spectrosc. **111**, 535 (2011)
27. Higginbottom, D.B., et al.: Pure single photons from a trapped atom source. New J. Phys. **18**, 093038 (2016)
28. Maurer, C., Becher, C., Russo, C., Eschner, J., Blatt, R.: A single-photon source based on a single ca$^+$ ion. New J. Phys. **6**, 94 (2004)
29. Albrecht, R., et al.: Narrow-band single photon emission at room temperature based on a single nitrogen-vacancy center coupled to an all-fiber-cavity. Appl. Phys. Lett. **105**, 073113 (2014)
30. Ou, Z.Y., Lu, Y.J.: Cavity enhanced spontaneous parametric down-conversion for the prolon-gation of correlation time between conjugate photons. Phys. Rev. Lett. **83**, 2556–2559 (1999)
31. Scholz, M., Koch, L., Benson, O.: Statistics of narrow-band single photons for quantum mem-ories generated by ultrabright cavity-enhanced parametric down-conversion. Phys. Rev. Lett. **102**, 063603 (2009)
32. Zhang, H., et al.: Preparation and storage of frequency-uncorrelated entangled photons from cavity-enhanced spontaneous parametric downconversion. Nat. Photonics **5**, 628–632 (2011)
33. Fekete, J., Rieländer, D., Cristiani, M., de Riedmatten, H.: Ultranarrow-band photon-pair source compatible with solid state quantum memories and telecommunication networks. Phys. Rev. Lett. **110**, 220502 (2013)
34. Rambach, M., Nikolova, A., Weinhold, T.J., White, A.G.: Sub-megahertz linewidth single photon source. APL Photonics **1** (2016)
35. Brown, R.H., Twiss, R.Q.: Correlation between photons in two coherent beams of light. Nature **177**, 27–29 (1956)

36. Hong, C.K., Ou, Z.Y., Mandel, L.: Measurement of subpicosecond time intervals between two photons by interference. Phys. Rev. Lett. **59**, 2044–2046 (1987)
37. Glauber, R.J.: The quantum theory of optical coherence. Phys. Rev. **130**, 2529–2539 (1963)
38. Loudon, R.: The Quantum Theory of Light, 1st edn. Clarendon Press, Oxford (1973)
39. Fasel, S., et al.: High-quality asynchronous heralded single-photon source at telecom wavelength. New J. Phys. **6**, 163 (2004)
40. Grangier, P., Roger, G., Aspect, A.: Experimental evidence for a photon anticorrelation effect on a beam splitter: a new light on single-photon interferences. Europhys. Lett. **1**, 173 (1986)
41. Kwiat, P.G., Chiao, R.Y.: Observation of a nonclassical berry's phase for the photon. Phys. Rev. Lett. **66**, 588–591 (1991)
42. Poh, H.S., Joshi, S.K., Cerè, A., Cabello, A., Kurtsiefer, C.: Approaching tsirelson's bound in a photon pair experiment. Phys. Rev. Lett. **115**, 180408 (2015)
43. Knill, E., Laflamme, R., Milburn, G.J.: A scheme for efficient quantum computation with linear optics. Nature **409**, 46–52 (2001)
44. Clausen, C., et al.: Quantum storage of photonic entanglement in a crystal. Nature **469**, 508–511 (2011)
45. Lu, Y.J., Ou, Z.Y.: Optical parametric oscillator far below threshold: experiment versus theory. Phys. Rev. A **62**, 033804 (2000)
46. Kuklewicz, C.E., Wong, F.N.C., Shapiro, J.H.: Time-bin-modulated biphotons from cavity-enhanced down-conversion. Phys. Rev. Lett. **97**, 223601 (2006)
47. Neergaard-Nielsen, J.S., Nielsen, B.M., Takahashi, H., Vistnes, A.I., Polzik, E.S.: High purity bright single photon source. Opt. Express **15**, 7940–7949 (2007)
48. Bao, X.-H., et al.: Generation of narrow-band polarization-entangled photon pairs for atomic quantum memories. Phys. Rev. Lett. **101**, 190501 (2008)
49. Wang, F.-Y., Shi, B.-S., Guo, G.-C.: Generation of narrow-band photon pairs for quantum memory. Opt. Commun. **283**, 2974–2977 (2010)
50. Chuu, C.-S., Harris, S.E.: Ultrabright backward-wave biphoton source. Phys. Rev. A **83**, 061803 (2011)
51. Pomarico, E., Sanguinetti, B., Osorio, C.I., Herrmann, H., Thew, R.T.: Engineering integrated pure narrow-band photon sources. New J. Phys. **14**, 033008 (2012)
52. Zhou, Z.-Y., Ding, D.-S., Li, Y., Wang, F.-Y., Shi, B.-S.: Cavity-enhanced bright photon pairs at telecom wavelengths with a triple-resonance configuration. J. Opt. Soc. Am. B **31**, 128–134 (2014)
53. Monteiro, F., Martin, A., Sanguinetti, B., Zbinden, H., Thew, R.T.: Narrowband photon pair source for quantum networks. Opt. Express **22**, 4371–4378 (2014)
54. Luo, K.-H., et al.: Direct generation of genuine single-longitudinal-mode narrowband photon pairs. New J. Phys. **17**, 073039 (2015)
55. Förtsch, M., et al.: Highly efficient generation of single-mode photon pairs from a crystalline whispering-gallery-mode resonator source. Phys. Rev. A **91**, 023812 (2015)
56. Ahlrichs, A., Benson, O.: Bright source of indistinguishable photons based on cavity-enhanced parametric down-conversion utilizing the cluster effect. Appl. Phys. Lett. **108**, 021111 (2016)
57. Sparkes, B.M., Hosseini, M., Hétet, G., Lam, P.K., Buchler, B.C.: An AC stark gradient echo memory in cold atoms. Phys. Rev. A **82**, 043847 (2010)
58. Sparkes, B.M., et al.: Gradient echo memory in an ultra-high optical depth cold atomic ensemble. New J. Phys. **15**, 085027 (2013)
59. Higginbottom, D.B., et al.: Spatial-mode storage in a gradient-echo memory. Phys. Rev. A **86**, 023801 (2012)
60. Cho, Y.-W., et al.: Highly efficient optical quantum memory with long coherence time in cold atoms. Optica **3**, 100–107 (2016)
61. Steck, D.A.: Rubidium 87 D line data (2015). http://steck.us/alkalidata/
62. Quantum States of Light. Springer, Berlin, Heidelberg (2015)
63. Träger, F.: Handbook of Lasers and Optics. Springer Handbooks, 1st edn. Springer, Berlin, Heidelberg (2007)

64. Cregan, R.F., et al.: Single-mode photonic band gap guidance of light in air. Science **285**, 1537 (1999)
65. Benabid, F., Couny, F., Knight, J., Birks, T., Russell, P.: Compact, stable and efficient all-fibre gas cells using hollow-core photonic crystal fibres. Nature **434**, 488–491 (2005)
66. Perrella, C., et al.: High-efficiency cross-phase modulation in a gas-filled waveguide. Phys. Rev. A **88**, 013819 (2013)
67. Saha, K., Venkataraman, V., Londero, P., Gaeta, A.L.: Enhanced two-photon absorption in a hollow-core photonic-band-gap fiber. Phys. Rev. A **83**, 033833 (2011)
68. Slepkov, A.D., Bhagwat, A.R., Venkataraman, V., Londero, P., Gaeta, A.L.: Spectroscopy of Rb atoms in hollow-core fibers. Phys. Rev. A **81**, 053825 (2010)
69. Perrella, C., Light, P.S., Stace, T.M., Benabid, F., Luiten, A.N.: High-resolution optical spectroscopy in a hollow-core photonic crystal fiber. Phys. Rev. A **85**, 012518 (2012)
70. Perrella, C., et al.: High-resolution two-photon spectroscopy of rubidium within a confined geometry. Phys. Rev. A **87**, 013818 (2013)
71. Okaba, S., et al.: Lamb-dicke spectroscopy of atoms in a hollow-core photonic crystal fibre. Nat. Commun. **5**, 4096 (2014)
72. Milburn, G.J.: Quantum optical fredkin gate. Phys. Rev. Lett. **62**, 2124–2127 (1989)
73. Araújo, M., et al.: Witnessing causal nonseparability. New J. Phys. **17**, 102001 (2015)
74. Chiribella, G., D'Ariano, G.M., Perinotti, P., Valiron, B.: Quantum computations without definite causal structure. Phys. Rev. A **88**, 022318 (2013)
75. Collett, M.J., Gardiner, C.W.: Squeezing of intracavity and traveling-wave light fields produced in parametric amplification. Phys. Rev. A **30**, 1386–1391 (1984)
76. Yang, J., et al.: Experimental quantum teleportation and multiphoton entanglement via interfering narrowband photon sources. Phys. Rev. A **80**, 042321 (2009)
77. Tian, L., Li, S., Yuan, H., Wang, H.: Generation of narrow-band polarization-entangled photon pairs at a rubidium D1 line. J. Phys. Soc. Jpn. **85**, 124403 (2016)
78. Scholz, M., Koch, L., Ullmann, R., Benson, O.: Single-mode operation of a high-brightness narrow-band single-photon source. Appl. Phys. Lett. **94** (2009)
79. Scholz, M., Koch, L., Benson, O.: Analytical treatment of spectral properties and signal-idler intensity correlations for a double-resonant optical parametric oscillator far below threshold. Opt. Commun. **282**, 3518–3523 (2009)
80. Schunk, G., et al.: Interfacing transitions of different alkali atoms and telecom bands using one narrowband photon pair source. Optica **2**, 773–778 (2015)
81. Schunk, G., et al.: Frequency tuning of single photons from a whispering-gallery mode resonator to MHz-wide transitions. J. Mod. Opt. **63**, 2058–2073 (2016)
82. Wolfgramm, F., et al.: Bright filter-free source of indistinguishable photon pairs. Opt. Express **16**, 18145–18151 (2008)
83. Haase, A., Piro, N., Eschner, J., Mitchell, M.W.: Tunable narrowband entangled photon pair source for resonant single-photon single-atom interaction. Opt. Lett. **34**, 55–57 (2009)
84. Wolfgramm, F., de Icaza Astiz, Y.A., Beduini, F.A., Cerè, A., Mitchell, M.W.: Atom-resonant heralded single photons by interaction-free measurement. Phys. Rev. Lett. **106**, 053602 (2011)
85. Rieländer, D., Lenhard, A., Mazzera, M., de Riedmatten, H.: Cavity enhanced telecom heralded single photons for spin-wave solid state quantum memories. New J. Phys. **18**, 123013 (2016)
86. de Riedmatten, H., Afzelius, M., Staudt, M.U., Simon, C., Gisin, N.: A solid-state light-matter interface at the single-photon level. Nature **456**, 773–777 (2008)
87. de Riedmatten, H.: Quantum optics: light storage at record bandwidths. Nat. Photonics **4**, 206–207 (2010)
88. Chuu, C.-S., Yin, G.Y., Harris, S.E.: A miniature ultrabright source of temporally long, narrowband biphotons. Appl. Phys. Lett. **101**, 051108 (2012)
89. Loredo, J.C., et al.: Scalable performance in solid-state single-photon sources. Optica **3**, 433–440 (2016)
90. Lu, Y.J., Campbell, R.L., Ou, Z.Y.: Mode-locked two-photon states. Phys. Rev. Lett. **91**, 163602 (2003)
91. Xie, Z., et al.: Harnessing high-dimensional hyperentanglement through a biphoton frequency comb. Nat. Photonics **9**, 536–542 (2015)

Chapter 2
Theoretical and Experimental Foundations

This chapter explains and summarises the theoretical concepts and experimental techniques necessary to understand the following chapters and includes various references for deeper insight. Sections 2.1 and 2.2 discuss the basics of optical cavities and introduce important techniques of frequency stabilisation. The focus in these parts lies on design considerations and mode-matching for cavities as well as the Pound-Drever-Hall technique [1] to stabilise the emission spectrum of lasers and cavities. Next, Sect. 2.3 presents the theory of second order nonlinear processes, aiming at a profound comprehension of second harmonic generation (SHG) and spontaneous parametric down-conversion (SPDC) to generate ultraviolet pump light and single photon pairs, respectively. It also introduces the heart of our experiment: optical parametric oscillators. Finally, Sect. 2.4 presents the derivation of crucial metrics of single photon sources. The section introduces classical and quantum mechanical tools to characterise the spectral brightness, linewidth, multi-photon suppression and indistinguishability of a single photon source.

2.1 Optical Cavities

Optical cavities are resonators consisting of a set of two or more mirrors between which the light circulates on a closed path. They are the fundamental building blocks of many applications. In lasers, cavities combined with a gain medium are used to build up optical power [2]. They can function as highly sensitive sensors like in gravitational wave detecting interferometers [3], measure low-level losses, for example in ring-down spectroscopy [4] or simply filter a frequency or spatial component of light. Throughout this thesis, carefully designed cavities are used to enhance the emission and manipulate the output spectrum of nonlinear crystals. The following sections review the basics to achieve this.

© Springer Nature Switzerland AG 2018
M. Rambach, *Narrowband Single Photons for Light-Matter Interfaces*,
Springer Theses, https://doi.org/10.1007/978-3-319-97154-4_2

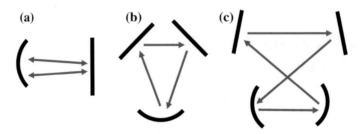

Fig. 2.1 Common optical cavity geometries. (**a**) Hemispherical cavity, example of a standing wave cavity. Ring cavities: (**b**) Triangular cavity, consisting of two plane and one curved mirror. (**c**) Bow-tie cavity: two curved and two plane mirrors

2.1.1 Resonator Theory

There are two basic types of resonators: a standing wave resonator and a ring resonator. The simplest version of a standing wave resonator consists of two mirrors separated by a distance $L = L_{rt}/2$, with L_{rt} the total length of one round trip. A ring resonator on the other hand includes a minimum of three mirrors, with usually at least one of them curved. Figure 2.1 depicts various common cavity configurations.

Light travelling between mirrors forms a standing wave if the wavelength λ of the light is an integer multiple of the resonator length, $L_{rt} = n\lambda$, with $n \in \mathbb{N}$. This relation combined with the boundary conditions at the mirror surfaces leads to the definition of the free spectral range (FSR), ν_{FSR}, the frequency spacing between two adjacent cavity modes:

$$\nu_{FSR} = \frac{c}{L_{rt}} = \frac{1}{t_{rt}}, \tag{2.1}$$

where c is the speed of light and t_{rt} is the round-trip time of the cavity. The frequency modes have a Lorentzian shape for low loss cavities with a full width half maximum linewidth of

$$\Delta\nu = \frac{1}{t_{rd}}. \tag{2.2}$$

Here, t_{rd}, the ring down time of the cavity after which the intensity of the circulating field decayed to $1/e$ of the steady state intensity. The ring down time depends only on the losses per round-trip inside the cavity and together with the round trip time we can define the finesse [2, 5]:

$$\mathcal{F} = \frac{t_{rd}}{t_{rt}} = \frac{\nu_{FSR}}{\Delta\nu} = \frac{\pi\sqrt{g_{rt}}}{1 - g_{rt}}, \tag{2.3}$$

where

$$g_{rt} = \sqrt{R_1 R_2 ... (1 - P_{loss})} = \sqrt{R_{tot}}. \tag{2.4}$$

Here, g_{rt} (always <1 for a passive cavity) is the cavity round-trip gain, the fraction of the power left after one round-trip. R_i is the reflectivity of mirror i and P_{loss} accounts for absorption or any additional losses on extra surfaces. The finesse can be interpreted as the average number of round trips of a photon before leaving the cavity through the outcoupling mirror or before being scattered and lost. It is a convenient number to characterise the quality of cavities, whereas the FSR and the linewidth fully characterise the emission spectrum. For a typical cavity used to produce the results of this thesis, one round trip is roughly 2.5 m, resulting in a FSR around 125 MHz. The finesse is around 200, corresponding to a round-trip gain of 96.5% and a linewidth of approximately 650 kHz, but more on that later.

Now that we defined the basic characteristics of a cavity we will have a look at the reflected, transmitted and circulating intensity of a cavity on resonance. The circulating intensity inside the resonator can be written simply as

$$I_{circ} = bI_0, \tag{2.5}$$

with the amplification factor [2]

$$b = \frac{T_{in}\mathcal{F}^2}{\pi^2 g_{rt}}. \tag{2.6}$$

I_0 is the intensity of the incoming field and $T_{in/out}$ is the transmittivity of the incoupling or outcoupling mirror. We can see from Eq. 2.6 that the circulating power depends on the ratio of the incoupler transmittivity to the total gain and increases quadratically with the finesse. We can use those two parameters to increase the field inside the cavity and therefore the creation rate of single photon pairs described in the next sections. The formulas for the reflected and transmitted intensity are given by [2]:

$$I_{ref} = \frac{(R_{in} - g_{rt})^2}{R_{in}(1 - g_{rt})^2}I_0, \tag{2.7}$$

$$I_{trans} = \frac{T_{in}T_{out}}{(1 - g_{rt})^2}I_0. \tag{2.8}$$

Equations 2.5, 2.6 and 2.8 show a tradeoff between the circulating and transmitted intensity for the single photon creation. In order to achieve desirable high photon pair rates we need a high field amplitude inside the cavity. This can be achieved by low losses and therefore a high finesse but in order to have a high escape efficiency of the photons we need to increase the transmission of the outcoupling mirror, subsequently lowering the finesse and broadening the linewidth.

A closer look at Fig. 2.2 exhibits three regimes for the cavity coupling dependent on the reflectivity of the incoupling mirror. In the over-coupled case the reflectivity is too high and only a small amount of the input intensity is propagating into the cavity, limiting the sub-sequent build-up. The under-coupled regime on the other hand allows the light to enter the cavity but there will be less intensity built up inside

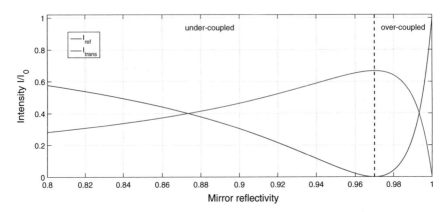

Fig. 2.2 Reflected and transmitted intensity as a function of the mirror reflectivity, with their extrema at $R = 0.97$, indicated by the dashed black line. I_{ref} is simulated with an total round-trip gain of $0.97 * R_{in}$ leading to $I_{ref}(0.97) = 0$. This means that the cavity is perfectly coupled and no power is reflected. I_{trans} is simulated with $T_{in} = 2\%$ and the same round-trip gain. We can see that the maximum is reached at the same reflectivity as before but not all power will be transmitted due to additional losses inside the cavity, i.e. I/I_0 does not reach 1. Perfect transmission can only be achieved if the incoupling and outcoupling mirror reflectivities are equal and are the only losses in the system following Eq. 2.8. The dashed line at critical coupling separates the under- (left) and over-coupled regime (right)

as the losses from the incoupling mirror dominate. In between the two lies the point of critical coupling (dashed line in Fig. 2.2), with low reflected intensity, high build-up and high transmission. If T_{in} ($= 1 - R_{in}$) equals the sum of all other losses in the resonator, perfect impedance matching is achieved as can be seen in Fig. 2.2. At this point no power is reflected off the cavity and the overall transmission reaches a maximum. The value of the transmission maximum depends on the losses inside the system: if T_{in} and T_{out} are equal and the only losses, 100% transmission is achieved.

2.1.2 Mode-Matching

The spatial modes of a cavity mainly depend on the shape of its mirrors. The most basic supported mode of a spherical mirror resonator is the Gaussian mode, usually referred to as the transverse electromagnetic zero-zero (TEM_{00}) mode. It is the fundamental solution of the Helmholtz equations with the appropriate boundary conditions. By definition, the transverse field distribution of a Gaussian beam follows a Gaussian profile. The intensity is a function of the axial and radial positions, z and $r = \sqrt{x^2 + y^2}$, respectively, and can be written as [5]

$$I(r, z) = I_0 \left[\frac{w_0}{w(z)} \right]^2 \exp \left[-\frac{2r^2}{w^2(z)} \right], \tag{2.9}$$

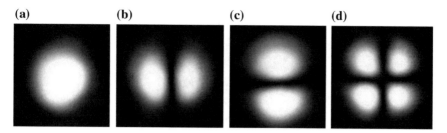

(a)   **(b)**   **(c)**   **(d)**

Fig. 2.3 Lowest four TEM_{nm} modes, pictures taken from one of our cavities. n and m describe the number of horizontal and vertical nodes, respectively. (**a**) TEM_{00}, fundamental Gaussian mode. (**b**) TEM_{10}, (**c**) TEM_{01} and (**d**) TEM_{11}

with

the waist at z = 0

$$w_0 = \sqrt{\frac{\lambda z_R}{\pi}}, \tag{2.10}$$

and every other position

$$w(z) = w_0 \sqrt{1 + \left(\frac{z}{z_R}\right)}, \tag{2.11}$$

The optical intensity is described by two independent parameters: I_0, the square of the electric field amplitude and z_R, the Rayleigh length where the wavefront curvature is at its maximum. Both parameters are determined by the boundary conditions of the cavity. All other parameters can be calculated if z_R and the wavelength λ in the medium of propagation are known. Equation 2.11 is referred to as the beam radius in the transverse plain, as $\left(1 - \frac{1}{e^2}\right) \approx 86\%$ of the power is within a circle of that radius. Its minimum value is the beam waist radius w_0, with the diameter $2w_0$ commonly called the waist spot size.

Different spatial modes can propagate stably inside a cavity. The modes are self-producing transverse intensity patterns and can be described exactly as Ince-Gaussian functions, with the two well-known special cases Hermite-Gaussian (HG) and Laguere-Gaussian (LG) functions for rectangular- and cylindrical-symmetric modes, respectively. Figure 2.3 shows the fundamental and the lowest three higher order HG modes, TEM_{nm}, of one of the cavities in our setup. Careful design of the laser coupling, beam waist and position of the waist to overlap with the desired fundamental cavity mode is called mode-matching. The goal of this is to use optical components to shape the incoming beam so that its wavefront will keep retracing itself while travelling between the cavity mirrors over multiple round-trips. This is, in general, not a trivial task. The standard technique to describe the propagation of a Gaussian beam through an optical system evolves the complex beam parameter q using the ABCD matrix formalism. A comprehensive introduction to this formalism can be found e.g. in Ref. [2]. The q-parameter is defined as

$$q(z) = z + iz_R, \tag{2.12}$$

$$\frac{1}{q(z)} = \frac{1}{R(z)} - i\frac{\lambda}{\pi w^2(z)}, \tag{2.13}$$

with the radius of curvature $R(z)$ of the wavefront:

$$R(z) = z\left[1 + \left(\frac{z_R}{z}\right)^2\right]. \tag{2.14}$$

Equation 2.12 describes the q-parameter intuitively from the solution of the Helmholtz equation whereas Eq. 2.13 separates the complex envelope of the Gaussian beam into an amplitude and a phase. Both representations are equivalent and have advantages in different situations. The ABCD formalism now relates the q-parameters q_1 and q_2 of a Gaussian beam before and after passing an optical system to each other via

$$q_2 = \frac{Aq_1 + B}{Cq_1 + D}. \tag{2.15}$$

(A, B, C, D) are elements of the matrix M that describes the optical system. This method can be used to calculate the waist(s) and waist position(s) of the spatial cavity mode as well as to tailor the incoming beam for optimal mode-matching.

We can now derive analytic expressions for the waists and their positions of a bow-tie cavity as in Fig. 2.1c using the ABCD formalism and comparing the results to a more elegant approach [6]. The cavity consists of four mirrors, two planar and two curved, between which the light propagates. Every curved mirror effectively acts as a lens and will lead to a beam waist of the cavity mode at a designated position. To form a stable, self-producing standing wave, the q parameter at the starting point has to be the same as the one after passing the optical system. With $q = q_1 = q_2$, Eq. 2.15 simplifies to

$$q = \frac{Aq + B}{Cq + D}. \tag{2.16}$$

Solving 2.16 means solving a quadratic equation for q, with only one solution being of physical significance:

$$q = \frac{A - D}{2C} + i\frac{\sqrt{4 - (A + D)^2}}{2C}. \tag{2.17}$$

Comparing Eqs. 2.17–2.12 and inserting the solution into Eq. 2.10 results in the desired parameters, beam waist and its position:

$$w_0^2 = \frac{\lambda}{2\pi C}\sqrt{4 - (A + D)^2}, \tag{2.18}$$

$$z = \frac{A - D}{2C}. \tag{2.19}$$

We can draw an additional important conclusion from Eq. 2.18: as $w_o \in \mathbb{R}^+$,

$$4 - (A + D)^2 \geq 0, \tag{2.20}$$

$$|A + D| \leq 2. \tag{2.21}$$

Only optical cavities that fulfil the stability criterion (Eq. 2.21) can support modes for a large number of round-trips. Otherwise, the beam size will grow further and further, eventually become bigger than the mirror and the light is lost.

In order to derive both beam waists and their positions we need to calculate the ray matrices M_1 and M_2, one starting the optical path at the first curved mirror, the other starting at the second. The solutions for our system consisting of two mirrors with radius of curvature R, separated by distances L_1 and $L_2 = L_{rt} - L_1$ from each other are:

$$M_1 = M_R M_{L_2} M_R M_{L_1}$$
$$= \begin{bmatrix} 1 - \frac{2L_2}{R} & L_1 + L_2 - \frac{2L_1L_2}{R} \\ \frac{4(L_2-R)}{R^2} & \frac{4L_1L_2 - 4L_1R - 2L_2R + R^2}{R^2} \end{bmatrix}, \tag{2.22}$$

$$M_2 = M_R M_{L_1} M_R M_{L_2}$$
$$= \begin{bmatrix} 1 - \frac{2L_1}{R} & L_1 + L_2 - \frac{2L_1L_2}{R} \\ \frac{4(L_1-R)}{R^2} & \frac{4L_1L_2 - 4L_2R - 2L_1R + R^2}{R^2} \end{bmatrix}, \tag{2.23}$$

with M_R and M_{L_i} the ABCD matrices for the curved mirrors and free space propagation, respectively, given by

$$M_R = \begin{bmatrix} 1 & 0 \\ \frac{-2}{R} & 1 \end{bmatrix}, \tag{2.24}$$

$$M_{L_i} = \begin{bmatrix} 1 & L_i \\ 0 & 1 \end{bmatrix}. \tag{2.25}$$

The matrices $M_{1,2}$ lead to waist positions exactly in between the two curved and plane mirrors for a symmetric system. The analytical solutions are

$$w_{0,1}^2 = \frac{\lambda}{2\pi} \sqrt{\frac{R - L_1}{R - L_2}} \, (R\,(L_1 + L_2) - L_1 L_2), \tag{2.26}$$

$$w_{0,2}^2 = \frac{\lambda}{2\pi} \sqrt{\frac{R - L_2}{R - L_1}} \, (R\,(L_1 + L_2) - L_1 L_2), \tag{2.27}$$

$$z_1 = \frac{L_1}{2}, \tag{2.28}$$

$$z_2 = \frac{L_2}{2}. \tag{2.29}$$

These formulas perfectly reproduce the results from [6], where general analytic expressions for an arbitrary amount of focussing elements are derived. It is a great example of the power and simplicity of the ABCD matrix formalism. The same technique is also used to calculate the necessary optics for manipulation of the incoming laser beam to match the calculated cavity mode.

There are a number of steps to be followed to achieve good mode-matching in an actual experiment. After some rough adjustment of the size and the position of the incoming beam, a camera behind the cavity is usually used to monitor the overlap of different cavity round-trips. Multiple iterations of aligning the incoming beam, but also the cavity mirrors themselves, help to project both modes onto each other. After identifying the desired mode (TEM$_{00}$) on the camera, the signal of a fast photodetector (see Sect. 3.5.1) can be monitored on an oscilloscope while fine adjusting the cavity mirrors until the higher-order HG modes vanish and all incoming intensity is deployed in the fundamental Gaussian mode.

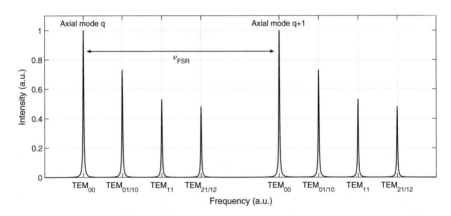

Fig. 2.4 Cavity transverse mode spectrum, normalised to 1 for the TEM$_{00}$ mode. The distance between adjacent axial modes is the free spectral range. Spacing and position of the higher order modes is dependent on the Gouy phase shift related to the cavity parameters and can therefore be engineered through the resonator. Mode-matching is aiming to suppress higher order TEM modes, deploying the incoming intensity solely into the fundamental TEM$_{00}$ mode

A typical transverse mode spectrum of a cavity before mode-matching is shown in Fig. 2.4. Spacing and position of the modes in the spectrum depend on the Gouy phase shift from one end of the cavity to the other, related to the Gaussian beam parameters and hence to the total resonator length, mirror curvature and spacing. The spectrum shows that individual higher order modes have different frequencies, but the dependency on cavity parameters generally allows to engineer the mode-structure of the resonator. For example, in a confocal two-mirror resonator all even-symmetry TEM modes are degenerate with the axial mode frequencies and all odd-symmetry modes overlap halfway between the axial mode locations [2]. Mode-matching now tries to couple the incoming light into the fundamental axial TEM_{00} modes while suppressing all others. More details on the design and mode-matching process of the cavity can be found in Sect. 3.1.

2.2 Stabilisation Techniques

Frequency stabilisation is one of the essential ingredients when building a narrow-band single photon source. The idea of active frequency control is the same for any reference system and generally quite simple: the frequency of a signal is measured with respect to a reference, e.g. the length of an optical cavity or the frequency of an atomic transition, and the derived error signal is fed back electronically to control the initial system. Figure 2.5 shows the schematic diagram of the process. The control circuit calculates the error signal out of the deviation between the initial and the reference frequency and then tries to compensate the difference. Experimentally implemented stabilisation cycles might consist of multiple feedback loops, serving different purposes: stabilisation to cavities allows designable linewidths, but only relative precision, while atomic transitions serve as steady, absolute frequency references but their tunability is poor.

There are a variety of different techniques to derive the error signal. Fringe side locking uses the side of a transmission fringe to convert frequency fluctuations into amplitude fluctuations with high efficiency [7], but low resolution, high drift rates and the sensitivity to power fluctuations make it unsuitable for certain applications. Additionally, the technique does not stabilise to transmission peaks and is therefore unfitting for our experiment. The Hänsch-Couillaud method [8] and its optimised version [9] utilise polarisation rotation to lock optical resonators without the need of phase modulation, but require an additional optical element inside the cavity. The locking cycles in this thesis operate with the Pound-Drever-Hall (PDH) technique [1], using the beat signal between the carrier frequency and modulation sidebands. This technique is fast, has a large noise cancellation bandwidth and is insensitive to power and spatial drifts of the laser. As PDH locking is a key technique in this thesis, it is more thoroughly discussed in the following section.

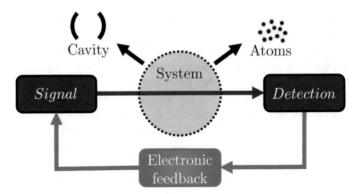

Fig. 2.5 Frequency feedback loop. The signal is referenced to a system and then detected. The electronic feedback creates an error signal out of the deviation between initial and reference frequency and then tries to eliminate the deviation. Throughout this thesis, the reference system is either the length of an optical cavity or the frequency of an atomic transition

2.2.1   Pound-Drever-Hall Frequency Stabilisation

The Pound-Drever-Hall (PDH) technique is a very robust frequency stabilisation method and reasonably easy to implement. It can achieve high noise cancellation bandwidths of several megahertz by modulating the frequency of the light. The resulting error signal has a high signal-to-noise ratio and odd symmetry around the central locking point, enabling stabilisation to the top of a transmission (or reflection) fringe. The main idea of the PDH technique is to phase modulate sidebands onto the carrier laser frequency and monitor the transmitted (reflected) signal. The sidebands allow straight forward detection of the antisymmetric derivative of the signal and the error signal can be derived by demodulating that signal at the sideband frequency.

The following paragraphs follow a review of the PDH technique [10] to understand the theory behind this simple but effective method. We start by defining the electric fields of a incoming laser beam $E_{in} = E_0 e^{i\omega t}$ and beam reflected of the cavity $E_{ref} = E_1 e^{i\omega t}$. Here, $E_{0,1}$ and ω are the complex amplitudes of the fields and the carrier frequency, respectively. It is important to note that the reflected beam is actually a coherent sum of two beams: the reflected beam that bounces off the incoupling mirror and a leakage beam, which is a small part of the standing wave inside the cavity that leaks backwards through the same mirror. The two beams have the same frequency but different relative phases dependent on the cavity resonance. At perfect resonance the two beams are π out of phase and annihilate each other. Slightly off resonance the phase difference is $\pi \pm \delta$, where the sign reveals on which side of the resonance the laser is sitting. One way to define and measure the phase relationship between the two beams is to modulate sidebands onto the laser carrier frequency (or phase) and detect the beat pattern at the modulation frequency, described in detail in Sect. 3.4.

Let us assume a sinusoidal phase modulation at a frequency Ω with a modulation depth β. The incoming electric field is now

$$E_{in} = E_0 e^{i[\omega t + \beta \sin(\Omega t)]}, \tag{2.30}$$

which can be rewritten in terms of Bessel functions $J_n(\beta)$:

$$E_{in} \approx E_0 \left[J_0(\beta) e^{i\omega t} + J_1(\beta) e^{i(\omega + \Omega)t} - J_1(\beta) e^{i(\omega - \Omega)t} \right]. \tag{2.31}$$

This is an approximation for small modulation depth ($\beta < 1$) and neglects all higher order sidebands J_n ($n > 1$). In this form it is straight forward to see that there are three beams at different frequencies present: a carrier at frequency ω, a lower and a higher sideband at frequencies $\omega \pm \Omega$, respectively.

In practise, there are two different regimes of operation which determine the overall shape of the error signal: slow modulation, like in saturation spectroscopy, where the linewidth is broader than the modulation frequency ($\Omega < \Delta\nu$) and fast modulation like in our optical cavity systems with a much narrower linewidth than the modulation frequency ($\Omega < \Delta\nu$). The error signals ϵ in these regimes can be approximated by [10]

$$\epsilon_{slow} \approx 2\sqrt{P_c P_s} \frac{d|F(\omega)|^2}{d\omega} \Omega, \tag{2.32}$$

$$\epsilon_{fast} \approx -2\sqrt{P_c P_s} \, \text{Im} \left\{ F(\omega)F^*(\omega + \Omega) - F^*(\omega)F(\omega - \Omega) \right\}. \tag{2.33}$$

Here, P_c and P_s are the total powers in the carrier and the sidebands, respectively, and $F(\omega) = {}^{E_{ref}}/_{E_{in}}$ is the ratio of the reflected and the incoming field at frequency ω. Figure 2.6 shows ϵ_{slow} and ϵ_{fast} dependent on the laser frequency and their integrals, defining the trapping potentials. Both error signals have a zero-crossing at resonance ($\omega = \nu_{FSR}$) and are antisymmetric, perfectly suited for frequency stabilisation. A closer look at the figure shows that the slow modulation regime has a far smaller noise cancellation bandwidth and a far shallower trapping potential compared to the fast modulation regime. Additionally, the fast modulation exhibits two extra zero-crossings at the sideband frequencies which are not resolved in the slow modulation case. These crossings result in an even deeper trapping potential and subsequently a tighter lock to the desired frequency.

2.2.2 Reference Systems

The shape and locking properties of the PDH error signal depend strongly on the bandwidth of the reference system and the applied sideband modulation frequency as derived in the previous section. Throughout this thesis there are two major reference systems in place: optical cavities and atomic vapours. Resonators allow switching

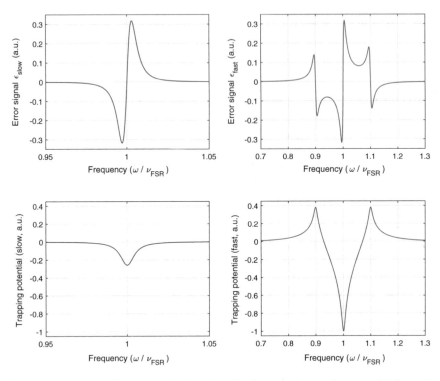

Fig. 2.6 Simulations of the error signal (top) and its integral, the trapping potential (bottom), dependent on incoming laser frequency for both regimes in PDH locking. On the left the resonance is broader than the modulation frequency, vice versa on the right. The round trip gain is 0.99 and the error signal strengths are scaled for easier comparison. Both error signals are antisymmetric and have a zero-crossing at resonance ($\omega = \nu_{FSR}$), well suited for frequency stabilisation. The trapping potential exhibits a much higher robustness and noise cancellation bandwidth of the lock on the right due to the higher ratio between modulation frequency and linewidth

between stabilisation of the their own length to a laser frequency, or vice versa, simply by changing the electronic feedback to either the laser or the cavity without any need to change the error signal. We already introduced different cavity geometries and calculated the PDH error signal detected on a beam reflected from the incoupling cavity mirror. This section introduces a way to stabilise lasers or cavities to a narrow transition of an atomic ensemble.

Atomic transitions are useful as a stable, well-characterised frequency reference that is largely independent of the environment. Therefore they can cancel out thermal fluctuations [11] leading to slow drifts of the system and precisely define the wavelength of the light. Although not straight forward to see, all previous derivations for the PDH technique are applicable in this case as well.

In general, the narrow hyperfine transitions of atoms predicted by quantum mechanics are not resolved in normal spectroscopy. At room temperature, thermal broadening of a transition line can be orders of magnitude wider than the desired spectroscopic features [12]: if an ensemble of atoms has a certain velocity distribution due to its temperature, the centre of the transition gets shifted for each atom individually, dependent on the direction and speed (Doppler effect), leading to a broadened line. However, using Doppler-free absorption spectroscopy [12, 13] and the PDH technique allows to derive an error signal from the hyperfine transitions without the need for cooling of the atoms.

The experimental setup is based on a pump-probe scheme to measure a transition frequency ω_0, shown in Fig. 2.7: a laser beam (frequency ω, wave vector k) is divided up unevenly on a beam splitter into a strong pump and a weak probe beam. Both beams propagate almost anti-collinearlly through an absorption cell and are detected afterwards. Scanning the frequency of a strong probe beam excites atoms moving at all possible velocities $v = (\omega-\omega_0)/k$, leading to a thermally broadened spectrum. The weak probe beam will also get absorbed by all moving atoms because of different Doppler shifts for pump and probe beam due to counter propagating paths. However, the stationary atoms with $v = 0$ are excited by the pump beam and cannot absorb the probe beam, resulting in a narrow dip in the probe absorption spectrum. This technique is often called spectral hole burning.

As derived in Sect. 2.2.1, the error signal in the PDH technique is a result of the coherent interference between the reflected field at the incoupling mirror and a leakage field from inside the cavity. If the role of the cavity is replaced by an atomic vapour, we can still identify two interfering electric fields: one field E_N travels through the cell as if there are no atoms present while the field E_D is produced by incident-field-induced dipole oscillations of the atoms [14]. It can be shown that the transmitted field of this system has exactly the same form as the reflected field in the cavity case, where E_N replaces the reflected field and E_D the leakage field from the cavity case [15]. All further statements and calculations now become analogous

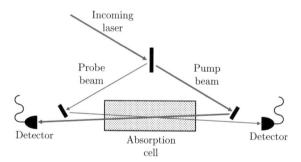

Fig. 2.7 Possible setup for Doppler-free absorption spectroscopy, adapted from [12]. An incoming laser beam gets split unevenly into a strong pump and a weak probe beam. Both enter an absorption cell from opposite directions, exciting atoms at different velocities due to the Doppler shift. Only stationary atoms see the individual beam frequencies unshifted and the strong pump beam excites these atoms while the probe beam passes undisturbed. The detected absorption spectrum has a narrow dip at the corresponding frequency which can be used to stabilise the laser via the PDH technique

to the cavity conditions, with the linewidth of the atomic transition playing the same role as the cavity linewidth before.

2.2.3 Proportional-Integral-Derivative Controller

As almost all branches of physics deal with unstable systems, stabilisation is very important. Laser frequency stabilisation, temperature control or length stabilisation of a cavity are just some examples relevant to this thesis. This section will give a brief introduction to the algorithms of control theory, especially proportional-integral-derivative (PID) control. More details can be found in the books [16, 17] or review articles, e.g. [18].

Proportional-integral-derivative (PID) control is probably the most commonly used control algorithm with a very robust performance and high simplicity in many applications. The goal of every feedback loop is to remove noise and environmental influences from the system. Ideally, the disturbances are canceled out fast and complete. The key feature of the PID controller with setpoint y_{sp} is the control signal $u(t)$, defined as [16]:

$$u(t) = k_p e(t) + k_i \int_0^t e(\tau)\, \mathrm{d}\tau + k_d \frac{\mathrm{d}}{\mathrm{d}t} e(t), \qquad (2.34)$$

where $e(t) = y_{sp} - y$ is the error signal with y, the measured variable. The control signal is a sum of three terms: a term purely proportional to the error signal (with gain k_p), a term proportional to the integral of the error signal (with gain k_i) and a term proportional to the derivative of the error signal (with gain k_d).

The P, I and D parts can be interpreted as actions based on the past, the present and the future of the system, respectively, and each of them fulfils a crucial task in the control loop. The P control provides a fast response and has the ability to partially cancel perturbations, but there will always be a residual error. The error can be decreased by increasing k_p, but this will also increase the probability of unwanted oscillations. However, adding the I control to the feedback will completely cancel out the steady state error from the P control. Higher gain will increase the strength of the I control, but again the tendency for oscillations increases accordingly. Consequently it is tempting to increase the P and I gain to get a fast and robust feedback loop, but at some point the system will start to oscillate. Intuitively, the D control can be seen as anticipating upcoming errors and taking immediate counteraction, which accelerates the response drastically. Combined with P and I, the D control allows for higher gains without unwanted oscillations, adding to the speed and robustness of the feedback loop. The problem with the D control lies in the assumptions it makes about the evolution of the system: in noisy systems with random fluctuations, the D part might add extra noise leading to an unstable system. Figure 2.8 shows an example of a P, a PI and a PID controlled system response and illustrates possible oscillations.

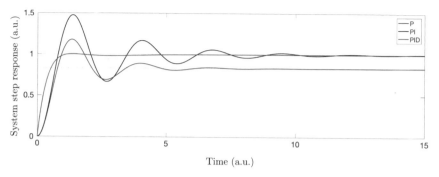

Fig. 2.8 Example of a feedback control with setpoint $y_{sp} = 1$. The blue line shows the system response to a pure P controller with $k_p = 5$: the response is fast but a steady-state error always remains. Adding an integral gain $k_i = 3$ (PI, black) slows down the response but cancels out any residual error. Finally, introducing an additional D control (PID with $k_d = 3$, red) allows for the fastest response of all three and compensates all disturbances

Table 2.1 Ziegler-Nichols method

Control	k_p (k_c)	k_i	k_d
P	0.50	–	–
PI	0.45	$0.54k_c/T_c$	–
PID	0.60	$1.20k_c/T_c$	$3k_cT_c/40$

As predicted, a carefully designed PID control has the fastest response and allows complete cancellation of any perturbations.

There are several methods to optimise a PID feedback system. It is crucial to always consider the key elements of your system: disturbance, sensor noise, uncertainties and reference signals. The different methods will not be described here but a heuristic tuning method, the Ziegler-Nichols method [16, 19], is introduced briefly. It is similar to a trial and error method where k_i and k_d are set to zero at the beginning and k_p is increased slowly until the system starts to oscillate. The gain and oscillation period at this critical point are defined as k_c and T_c, respectively. Then the P, I and D gain are adjusted as per Table 2.1.

These values apply for an ideal PID controller, but they are good starting points to further optimise realistic controllers. More details on the experimental implementation and optimisation of PID controls and the PDH technique are given in Sect. 3.4.

2.3 Nonlinear Optical Processes

The field of nonlinear optics deals with processes that occur when light interacts with matter and the response is nonlinearly dependent on the strength of the light field. This means that the refractive index n and the absorption coefficient α are not constant, but rather a complicated function of the intensity and therefore, in general, space, time, wavelength and polarisation. The light transmitted through a nonlinear medium can be completely different from the incoming light and produce new physical effects.

The field became important with the discovery of lasers, as the high intensities necessary to observe the nonlinear effects could only be achieved with laser light. The laser itself is probably the most prominent example of nonlinear optics. Since then, many nonlinear materials have been investigated for their properties in nearly every material system and various books [5, 20, 21] explain the topic in great detail. The following sections will introduce the background for general nonlinearities and focus on two second order (quadratic) effects, used in our system: second harmonic generation (SHG) and spontaneous parametric down-conversion (SPDC).

The response of a linear dielectric medium to an optical field $\mathbf{E}(\mathbf{r}, t)$ can be characterised by the polarisation $\mathbf{P}(\mathbf{r}, t)$:

$$\mathbf{P}(\mathbf{r}, t) = \epsilon_0 \chi(\mathbf{r}) \mathbf{E}(\mathbf{r}, t), \tag{2.35}$$

with ϵ_0, the vacuum permittivity and $\chi(\mathbf{r})$, the electric susceptibility of the medium. The susceptibility is a tensor and therefore the polarisation and the electric field not necessarily point into the same direction. It usually depends on the frequency and other properties of the light. In a nonlinear medium and for high intensities Eq. 2.35 can be expanded to

$$\mathbf{P}(\mathbf{r}, t) = \epsilon_0 \chi^{(1)}(\mathbf{r}) \mathbf{E}(\mathbf{r}, t) + \epsilon_0 \chi^{(2)}(\mathbf{r}) \mathbf{E}^2(\mathbf{r}, t) + \epsilon_0 \chi^{(3)}(\mathbf{r}) \mathbf{E}^3(\mathbf{r}, t) + \dots \tag{2.36}$$

The relation between polarisation and electric field is still linear for small $\mathbf{E}(\mathbf{r}, t)$, but nonlinear effects become dominant for high field strengths. Throughout the remainder of the thesis we will focus on the first two terms of Eq. 2.36. The susceptibilities are generally complex, leading to a change in phase and amplitude of the light field by refraction and absorption. In practise, $\chi^{(m)}$ are real tensors, sometimes only numbers, and therefore absorption is being neglected from here on.

The propagation of light in matter is described by the wave equation. In case of a homogeneous isotropic medium and neglecting higher order terms, the wave equation can be derived from the Maxwell equations and takes on the form [20]:

$$\nabla^2 \mathbf{E}(\mathbf{r}, t) - \frac{1}{c_0^2} \frac{d^2}{dt^2} (n_0 + \Delta n_{NL})^2 \, \mathbf{E}(\mathbf{r}, t) = 0. \tag{2.37}$$

Here, $\Delta n_{NL} = \sqrt{1 + \chi^{(2)}}$ is the changed refractive index due to the nonlinearity. The term $-\frac{\Delta n_{NL}^2}{c_0^2} \frac{d^2}{dt^2} \mathbf{E}(\mathbf{r}, t)$ describes a source of accelerated charges that emit electromagnetic radiation in a linear medium with refractive index $n_0 = \sqrt{1 + \chi^{(1)}}$. There are two approximate approaches to solving this nonlinear partial differential equation: the coupled wave theory and the Born approximation, using scattering theory [21].

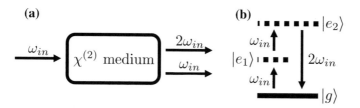

Fig. 2.9 (a) Schematic of the SHG process: the interaction of a plain wave with a medium creates an additional field component oscillating at double the frequency. (b) Energy level diagram describing SHG. Two photons excite the atomic ground state $|g\rangle$ to the virtual excited levels $|e_1\rangle$ and $|e_2\rangle$. $|e_2\rangle$ decays back to $|g\rangle$ by emitting a photon with frequency $2\omega_{in}$

2.3.1 Second Harmonic Generation

Second harmonic generation (SHG) is a nonlinear process that can be used to convert light from a lower to a higher frequency, e.g. infrared to visible or ultraviolet light. It is the simplest case of a second order nonlinear process where two equal monochromatic light waves, travelling along the z direction with the same frequency ω_{in}, wavevector k_{in} and polarisation, are simultaneously present in a nonlinear crystal. The nonlinear term of the atomic polarisation P_{NL} can then be written as [20]

$$
\begin{aligned}
P_{NL} &= \epsilon_0 \chi^{(2)} E_{in}^2 \\
&= \frac{1}{2}\epsilon_0 \chi^{(2)} E_0^2 + \frac{1}{2}\epsilon_0 \chi^{(2)} E_0^2 \cos(2\omega_{in}t) \\
&= P_{NL}(0) + P_{NL}(2\omega_{in}).
\end{aligned}
\tag{2.38}
$$

The first term in Eq. 2.38 is a steady component which describes a macroscopic charge separation inside the material. According to the previous section, we can interpret the second term as electromagnetic radiation oscillating at double the frequency of the incoming wave: $\omega_{SHG} = 2\omega_{in}$. This component of the optical field is called the second harmonic of the incoming field. As SHG fulfils energy and momentum conservation it can be used to generate light at twice the initial frequency that (almost) preserves the linewidth of the incoming light and travels along the same path. Figure 2.9 shows the schematic process (a) and the energy level diagram (b) of SHG. According to this picture, we can interpret SHG as an exchange of photons between different frequency components of the field, where creating a photon at $2\omega_{in}$ demands the annihilation of two photons at ω_{in}. The virtual excited states (dashed lines) are not energy levels of the atom, but rather describe combined energy levels of an atom with one or more photons of the light field [21]. It is important to point out that, in principle, SHG is also possible without atoms (in vacuum) [22], but for the experiments in this thesis the nonlinearities are introduced by atoms in crystals.

The amplitude of the SHG process is proportional to the intensity of the incoming field, consequently its intensity is proportional to the square of the incoming intensity.

The efficiency of the process, η_{SHG}, is dependent on the the incoming power P, the interaction cross section (or focus area) A and the crystal length l. It is defined as [5]

$$\eta_{SHG} \equiv \frac{I(2\omega)}{I(\omega)} = C^2 \frac{l^2}{A} P. \tag{2.39}$$

Here, C is a constant dependent on the crystal properties and the conversion frequency. η_{SHG} is below one, but under proper experimental conditions using long crystals, high powers and tight focussing, nearly all of the incident light can be transferred to double the frequency. The size of the laser beam and the length of the crystal can be engineered according to the application and are limited by the beam diffraction and the growing process of the crystal. The optimal focussing of the incident wave has been studied in great detail by Boyd and Kleinman [23]: the ratio of the crystal length l and the confocal parameter b of the laser beam should be $l/b \approx 2.84$. As the confocal parameter is two times the Rayleigh length z_R, we can use Eq. 2.10 to calculate the optimal beam waist $w_{0,opt}$ dependent on the wavelength in vacuum, the crystal length and refractive index:

$$w_{0,opt} = \sqrt{\frac{l\lambda_0}{2\pi \, 2.84 \, n}}, \tag{2.40}$$

known as the Boyd-Kleinman criterion. More recent investigations have shown that the restrictions on the beam waist are not as strict as previously assumed [24], leading to good conversion efficiencies even at $2w_{0,opt}$.

To achieve high powers necessary for high efficiencies, pulsed lasers with large peak powers are usually used. However, short pulses are inherently broad in frequency and, as SHG preserves the linewidth, would lead to pump light that is orders of magnitude broader than the down-conversion cavity linewidth (see Sect. 3.4.2). Additionally, a pulsed scheme would reduce the generation duty cycle, lowering the spectral brightness of the source. For continuous narrowband SHG at high powers, cavities are a great tool to create uninterrupted strong fields as discussed later.

The momentum conservation in nonlinear processes is not given automatically. A phase mismatch $\Delta k = k_{SHG} - 2k_{in}$ is usually present and can severely compromise the efficiency given in Eq. 2.39. Increasing the phase overlap of the incident and the newly generated light is called phase matching and can be achieved by choosing a suitable orientation of the crystal. All concepts for phase matching presented here are applicable for all frequency conversion techniques, especially for SPDC described in Sect. 2.3.3.

We can solve Eq. 2.37 in one dimension using the ansatz [20]:

$$E_{SHG}(z, t) = E_{0,SHG} \cos(\omega_{SHG} t - k_{SHG} z), \tag{2.41}$$

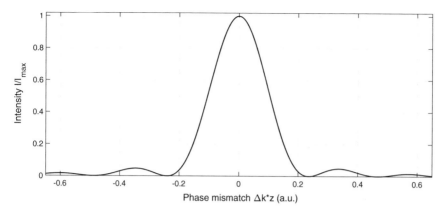

Fig. 2.10 Theoretical phase matching curve of a nonlinear crystal as a function of the mismatch Δkz. The curve exhibits the sinc^2 behaviour with maximum phase matching achieved for $\Delta kz = 0$. Away from optimum, a drastic decrease in output intensity can be observed

Some further calculations lead to an analytical expression for the SHG intensity

$$I_{SHG}(z) = I_{max}\left(\frac{\sin\left(\Delta kz/2\right)}{\Delta kz/2}\right)^2, \tag{2.42}$$

where

$$I_{max} = I_{in}^2 \frac{2\pi^2 d^2 l^2}{\epsilon_0 c_0 \lambda_{in}^2 n_{in}^2 n_{SHG}}, \tag{2.43}$$

with the different refractive indices n_i for the different wavelengths λ_i, the relevant matrix element d of the susceptibility and the length of the crystal l. Figure 2.10 illustrates the findings of Eq. 2.42. It shows a dramatic decrease in the SHG intensity for $\Delta kz \neq 0$, highlighting the importance of proper phase matching.

As pointed out before, momentum conservation ($\Delta k = k_{SHG} - 2k_{in} = 0$) together with energy conservation ($\omega_{SHG} - 2\omega_{in} = 0$) implies spatial phase matching. This is often difficult to achieve because nonlinear crystals show dispersion, so the refractive index is an increasing function of frequency. It is not possible to achieve momentum and energy conservation in such a material, but good phase matching can be obtained by carefully choosing the orientation and temperature of the crystal to equal the refractive index of the incident and frequency doubled wave. It is important to point out here that phase matching cannot be achieved for every crystal and some crystals might be preferable to others dependent on the wavelength and conversion type.

There are two main approaches to achieve phase matching: using birefringence and quasi-phase matching. Both require a high control of the refractive indices for all waves involved, usually achieved by additional angular and temperature tuning of the crystal. The birefringence method uses the dependence of the refractive index of a material on the polarisation of the optical wave. The light beam can be distinguished into ordinary (o) and extraordinary (e) wave with refractive indices n_o and n_e, respectively, dependent on the direction of polarisation. We can now identify two different types of phase matching. In type I, both incoming photons have the same polarisation, while the SHG wave is orthogonally polarised (oo-e, Eq. 2.44). In type II, the polarisations of the incident waves are oriented perpendicular to each other (eo-e, Eq. 2.45):

$$k_{SHG}(e) = k_{in}(o) + k_{in}(o), \tag{2.44}$$

$$k_{SHG}(e) = k_{in}(o) + k_{in}(e). \tag{2.45}$$

Both types have their advantages: type I allows not only spectral separation of the incident and converted light fields, but also distinction in polarisation. Type II on the other hand allows higher phase matching angles in suitable materials and in the case of SPDC, deterministic separation of the single photon pairs on a polarising beam splitter is possible. More on SPDC can be found in the upcoming section.

Angular and temperature tuning are the best options to manipulate the birefringence and fulfil Eqs. 2.44 and 2.45. A closer look at the angular dependence θ_{pm} of a type I phase matching process leads to [21]:

$$\sin^2 \theta_{pm} = \frac{n_{o,in}^{-2} - n_{o,SHG}^{-2}}{n_{e,SHG}^{-2} - n_{o,SHG}^{-2}}, \tag{2.46}$$

and usually reaches values of 40–60° dependent on the crystal. This equation determines how the crystal should be oriented to achieve optimal phase matching. Unfortunately, there is no analytical solution for a type II phase matching angle and the equation needs to be solved numerically.

However, there is one severe drawback of angular tuning: in practise, ordinary and extraordinary waves quickly diverge from each other. This walk-off limits the efficiency of the nonlinear process as the waves stop to overlap spatially. Temperature tuning can solve this problem by changing the amount of birefringence, which can be strongly dependent on the temperature, without adjusting the angle. Combining angular and temperature tuning results in less angular sensitivity and is called noncritical phase matching.

An alternative method, known as quasi-phase matching (QPM), can be used in cases where the birefringence technique is insufficient to fulfil $\Delta k = 0$, for instance for materials with low or no birefringence or when an application requires the use of a different element of the susceptibility tensor. Quasi-phase matching uses a periodic change (poling, see Sect. 2.3.2) in the nonlinearity of the crystal to reset the accumulated phase error before a back conversion starts, illustrated in red compared to dashed black in Fig. 2.11. The poling period Λ_{qpm} of the crystal depends on the natural phase mismatch of the light fields involved. For first order quasi-phase matching it is given as:

$$\Lambda_{qpm} = \frac{2\pi}{|\Delta k|} = \frac{2\pi}{|k_{SHG} - 2k_{in}|}. \tag{2.47}$$

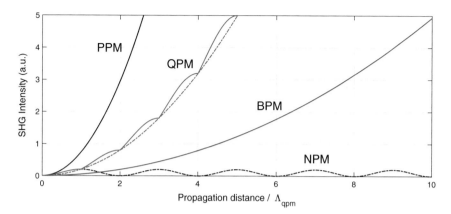

Fig. 2.11 Comparison of the field intensities dependent on the propagation distance through the crystal for different phase matching methods. The dashed black line shows the case of no phase matching (NPM): the intensity oscillates around a small value with period $2\Lambda_{qpm}$. The solid black line is perfect phase matching (PPM) with a quadratically increasing intensity. The quasi-phase matching (QPM) method is shown as solid and dashed red line for the actual intensity and its quadratical tendency, respectively. The blue line depicts the intensity in the birefringence phase matching (BPM) technique. The latter three all increase quadratically with different slopes proportional to d_{eff}

Typical poling periods are on the order of tens of micrometers for optical frequencies. Quasi-phase matching has two major advantages over the birefringence method: first, the polarisation of the interacting waves can be equal as QPM does not rely on birefringence. Second, QPM provides maximal design flexibility for many nonlinear processes at various interaction wavelengths.

Figure 2.11 compares the different phase matching approaches with perfect and no phase matching. In the case of no phase matching, the intensity oscillates around a small value with no significant total conversion intensity. All other cases show an increase in the intensity of the second harmonic field that depends quadratically on the distance travelled through the crystal, but the slopes differ significantly. The slopes are proportional to an effective susceptibility d_{eff} which is the highest for perfect phase matching and lowest for birefringence phase matching.

2.3.2 Periodically Poled Nonlinear Crystals

Periodic poling of nonlinear crystals is a well-established technique to achieve quasi-phase matching. It was first proposed in an article by Armstrong *et al.* in 1962 as a way to provide phase correction in SHG [25]. However, it took another 30 years of development to achieve SHG in a periodically poled (pp) crystal for the first time [26]. Growing high precision poling periods in nonlinear crystals with periods on the order of 10 micrometers for optical frequency conversion is still a challenge

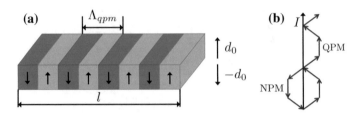

Fig. 2.12 (a) Nonlinear crystal with a poling period Λ_{qpm} to invert the nonlinear coefficient d_0 regularly. Figure adapted from Ref. [5]. (b) Build up of the intensity I with (QPM) and without poling (NPM) from amplitudes at different parts of the crystal. In the NPM case the phase misalignment leads to a circle without significant build up of intensity, whereas QPM periodically changes the sign of the nonlinearity, creating a high intensity conversion field

today. Figure 2.12 shows a first order pp crystal with poling period Λ_{qpm} for optimal quasi-phase matching in (a) and the intensity of the created waves in the not matched and QPM case (b) (see also Fig. 2.11).

In the upcoming experiments, periodically poled potassium titanyl phosphate (ppKTP) is used for SHG and SPDC. It has a high nonlinear optical coefficient $d_0 \approx 15$ pm/V and is transparent for wavelengths between 350 nm and 4.3 μm. The poling period for our application can be calculated using Eq. 2.57. To determine the wavevectors for the frequencies involve in the conversion process, precise knowledge of the relevant refractive indices is necessary. They are given by the phenomenological Sellmeier equations for the three different axis of ppKTP [27]:

$$n_x^2 = 3.29100 + \frac{0.04140}{\lambda^2 - 0.03978} + \frac{9.35522}{\lambda^2 - 31.45571}, \tag{2.48}$$

$$n_y^2 = 3.45018 + \frac{0.04341}{\lambda^2 - 0.04597} + \frac{16.98825}{\lambda^2 - 39.43799}, \tag{2.49}$$

$$n_z^2 = 4.59423 + \frac{0.06206}{\lambda^2 - 0.04763} + \frac{110.80672}{\lambda^2 - 86.12171}, \tag{2.50}$$

with λ in μm. The temperature dependence of the refractive indices for ppKTP is very small and was therefore neglected in the following calculations. For our application, n_z and n_y are engineered as the ordinary and extraordinary refractive indices, respectively. The poling period for conversion of light with a vacuum wavelength of roughly 397.5 to 795 nm (SPDC) and vice versa (SHG) is $\Lambda_{qpm} = 8.8$ μm. Periods of that magnitude can be manufactured by ferroelectric field poling, where electric fields induce the change in the crystal domain structure under carefully controlled conditions.

In the case of SPDC, we can calculate the bandwidth of the resulting photons by setting $\Delta kl = 2.7831$. The phase at this point results in only half of the frequency conversion compared to the centre. The corresponding difference between

$\Omega(\Delta k = {}^{2.7831}/_l)$ and $\Omega(\Delta k = {}^{-2.7831}/_l)$ around ω_0 is called the full width half maximum (FWHM) phase matching bandwidth and can be expressed as [28]:

$$\Delta\Omega = \frac{2\pi}{l|k'_s - k'_i|}, \qquad (2.51)$$

with $k'_{s,i} = \frac{dk_{s,i}}{d\omega_{s,i}}$, the derivative of the k-vectors evaluated at ω_0. For our case of degenerate SPDC, producing two single photons at 795 nm inside a 25 mm long crystal, the bandwidth is 100 GHz.

2.3.3 Spontaneous Parametric Down-Conversion

Spontaneous parametric down-conversion (SPDC) is roughly the inverse process of SHG. It creates single photon pairs rather than high intensity coherent fields and is therefore perfectly suited for optical quantum information and communication. SPDC is a special case of an optical parametric amplifier (see e.g. [5, 20, 21]) where a pump photon at frequency ω_p enters a suitable nonlinear medium and creates two photons, typically called signal and idler for historical reasons, at frequencies ω_S and ω_I, respectively. The term down-conversion relates to the created frequencies being lower than the pump frequency. Generally, signal and idler photons are correlated in their spatial, temporal, spectral and polarisation properties as shown below.

Unfortunately, SPDC is a very rare process and only happens with a probability on the order of 10^{-7}. Additionally, the resultant new photons are not as spectrally narrow as the pump photon due to the noisy nature of the process. The linewidth of unfiltered photons from SPDC is around 100 GHz to 1 THz, 5–6 orders of magnitude larger than typical bandwidths of atomic transitions. This section will discuss the nature of SPDC and its variations from SHG in detail.

The schematic SPDC process is depicted in Fig. 2.13. Since the decay at ω_I stimulates a parametric decay at ω_S and vice versa, the process is called parametric amplification [21]. The most dramatic difference between SHG and SPDC can be observed when comparing Figs. 2.13b–2.9b. In SHG, two photons add up to one single photon at double the frequency, roughly preserving the linewidth of the incoming field. The virtual excited energy levels are well defined by the energy of the two pump photons. In SPDC on the other hand, the input decays into two parts, but theoretically, the virtual state $|e_1\rangle$ can be anywhere between $|e_2\rangle$ and $|g\rangle$ as long as energy conservation is fulfilled. This leads to a widening of the down-converted linewidth compared to the pump field. Using quantum mechanics to describe the process further explains its spontaneous nature: SPDC can be understood as a special case of three-wave mixing [5], where one high-frequency photon can couple to the vacuum field and then randomly decays into two low-frequency photons.

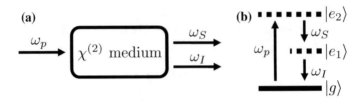

Fig. 2.13 (a) Classical schematic of the SPDC process: a pump field interacts with a nonlinear medium to create two photons at frequencies ω_S and ω_I. (b) Energy level diagram describing SPDC. A pump photon excites the atomic ground state $|g\rangle$ to the virtual excited level $|e_2\rangle$ from where it decays back down to $|g\rangle$ via $|e_1\rangle$, emitting two photons called signal and idler

Describing SPDC in a more mathematical way analogous to SHG, the intensity of the process is given by a sinc^2 function (see Eq. 2.42) dependent on the phase mismatch $\Delta\mathbf{k}$ with the phase matching condition

$$\Delta\mathbf{k} = \mathbf{k}_p - \mathbf{k}_S - \mathbf{k}_I = 0 \tag{2.52}$$

and energy conservation

$$\omega_p = \omega_S + \omega_I. \tag{2.53}$$

The case where $\omega_S = \omega_I = {}^{\omega_p}/_2$ is called frequency degenerate SPDC. Equations 2.52 and 2.53 lead to a strong correlation between signal and idler frequency if the pump spectrum is narrow. Like in the SHG case, we have type I and type II SPDC producing mutually parallel or orthogonally polarised photon pairs. As the wavevectors for signal and idler are generally different and distinguishable in frequency and/or polarisation, Eq. 2.45 transforms to Eqs. 2.55 and 2.56:

$$\text{Type I} \qquad \mathbf{k}_p(e) = \mathbf{k}_S(o) + \mathbf{k}_I(o), \tag{2.54}$$

$$\text{Type II} \qquad \begin{aligned} \mathbf{k}_p(e) &= \mathbf{k}_S(o) + \mathbf{k}_I(e), \tag{2.55} \\ \mathbf{k}_p(e) &= \mathbf{k}_S(e) + \mathbf{k}_I(o). \tag{2.56} \end{aligned}$$

Both phase matching approaches discussed in the Sect. 2.3.1 are also used in SPDC. However, the poling period for quasi-phase matching differs slightly from the SHG case because signal and idler are in general not identical in frequency and therefore both photons see different refractive indices which are themselves functions of frequency and temperature:

$$\Lambda_{qpm}(\text{T}) = \frac{2\pi}{|\Delta k|} = \frac{2\pi}{|k_p(\omega_p, \text{T}) - k_S(\omega_S, \text{T}) - k_I(\omega_I, \text{T})|}. \tag{2.57}$$

We can analyse the output spectrum by looking at small deviations of signal and idler frequency around optimal phase matching for a crystal with fixed length l and poling period Λ_{qpm}. This leads to the same sinc^2 behaviour as shown in Fig. 2.10 but in the frequency domain [28].

Finally, in order to use SPDC for quantum information or communication, a quantum mechanical treatment is necessary. Here, SPDC is described as fields generated by interactions with quantum vacuum fields with the following Hamiltonian:

$$H = g^* \, \hat{a}_p^\dagger \hat{a}_S \hat{a}_I + g \, \hat{a}_p \hat{a}_S^\dagger \hat{a}_I^\dagger, \qquad (2.58)$$

where g is the coupling constant describing the susceptibility of the crystal and $\hat{a}^\dagger$ and $\hat{a}$ are creation and annihilation operators of the respective fields, assuming a two-mode output of the SPDC. Without loss of generality, we can assume a strong pump field that can be treated classically and therefore Eq. 2.59 reduces to

$$H = g'^* \, \hat{a}_S \hat{a}_I + g' \, \hat{a}_S^\dagger \hat{a}_I^\dagger, \qquad (2.59)$$

with the classical pump field amplitude absorbed in g'. A more detailed discussion of the quantum nature of SPDC can be found in Refs. [29, 30]. The most important classical and quantum metrics of single photon sources are derived in Sect. 2.4.

2.3.4 Cavity-Enhanced Down-Conversion: Optical Parametric Oscillators

As pointed out in Sects. 2.3.2 and 2.3.3, the free running bandwidth of down-conversion ($\sim$100 GHz) is around 5 orders of magnitude too high for an efficient interaction with atomic transitions. Passive filtering with narrow-band optical filters or cavities might seem like the obvious choice to create suitable photons, but these methods turn out impractical due to the high losses involved in the process resulting in a low spectral brightness. A different approach is to create the single photons inside an optical cavity. In doing so we benefit from enhanced emission into a desired mode and map the spectral properties, especially the linewidth, from the cavity onto the single photons [31–44]. This feedback SPDC setup is also called an optical parametric oscillator (OPO). They have proven to be a flexible sources of radiation at many optical frequencies and can operate in continuous or pulsed mode.

We have previously shown that the three waves in SPDC have to fulfil frequency and phase matching conditions (Eqs. 2.52 and 2.53). If the photons are produced in a cavity, their frequencies must also overlap with the resonance frequencies of the cavity modes, similar to a laser amplifier. All other frequencies within the phase matching envelope interfere destructively. This leads to a modified spectrum as shown in Fig. 2.14: the phase matching curve is sampled by the narrowband modes of the cavity separated by the FSR. Another advantage of OPOs, apart from the narrowband

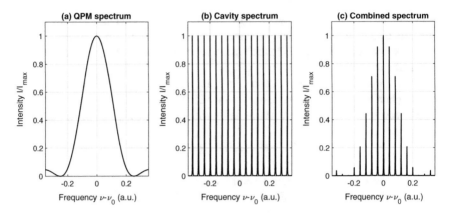

Fig. 2.14 (a) SPDC spectrum of a QPM crystal showing the sinc^2 dependency with a bandwidth of Ω. (b) Cavity spectrum consisting of Lorentzian peaks with width $\Delta\nu$, separated by $\nu_{FSR} \ll \Omega$. (c) Combined spectrum of a nonlinear crystal inside a cavity

emission, is that the effective length of the crystal is increased by the finesse of the cavity, greatly enhancing the spectral brightness. The intuition behind this lies within the interpretation of the finesse: it describes the average number of cavity round-trips of a photon. Thus, each pump photon transverses the crystal multiple times increasing its effective length compared to a single pass system. Finally, designing the linewidth of the cavity to match the atomic transition will produce single photon pairs that are resonant with the desired atomic species (see Sect. 3.1).

The threshold pump power for oscillation of an OPO is reached when the round-trip losses are less than the parametric gain, the increase in the converted field strength, resulting in a high conversion efficiency of the pump light into signal and idler beams. There are naturally three different options of resonance for an OPO: if only the signal is resonant, we speak of a singly resonant OPO. If signal and idler or signal and pump are resonant simultaneously, the OPO is doubly resonant or pump-enhanced singly resonant, respectively. In the third alternative, triple resonance of signal, idler and pump is achieved at the same time. The latter is the most challenging setup, often requiring complicated active stabilisation techniques. On the other hand triply resonant OPOs allow continuous wave operation with very low pump powers, making them our choice for the setup. The formula for the threshold photon flux N_t of a triply resonant OPO is [45, 46]:

$$N_t = \frac{(T_p + L_p)^2 (T_S + L_S)(T_I + L_I)}{64 T_p \chi_{eff}}, \qquad (2.60)$$

where $T_p, L_p, T_S, L_S, T_I, L_I$ are the transmission and loss coefficients for pump, signal and idler beams, respectively, and χ_{eff} is the effective nonlinear coefficient also taking the geometrical overlap of the three waves into account.

Dependent on the incoming photon flux (or pump power), we can identify three different regimes of operation of an OPO. Above threshold, the oscillation greatly enhances the nonlinearity leading to a coherent output of light that can be treated classically [21, 47]. Right below threshold, the OPO produces squeezed states of light [48–51]. The region that we are interested in is far below threshold, where we can neglect stimulated emission leading to higher order terms involving more than one photon pair. Here, the photon statistics are quantum.

In their groundbreaking work on cavity-enhanced SPDC, Ou and Lu [31] showed that, far below threshold, spontaneous emission is dominant for two-photon generation and stimulated emission is negligible. The output operator of a degenerate OPO at resonance is given by [48]:

$$\hat{a}_{out}(\omega_0 + \omega) = G_1(\omega)\hat{a}_{in}(\omega_0 + \omega) + g_1(\omega)\hat{a}^\dagger_{in}(\omega_0 - \omega) + G_2(\omega)\hat{b}_{in}(\omega_0 + \omega) + g_2(\omega)\hat{b}^\dagger_{in}(\omega_0 - \omega),$$
$$(2.61)$$

with

$$G_1(\omega) = \frac{\gamma_1 - \gamma_2 + 2i\omega}{\gamma_1 + \gamma_2 - 2i\omega},$$

$$g_1(\omega) = \frac{4\epsilon\gamma_1}{(\gamma_1 + \gamma_2 - 2i\omega)^2},$$

$$G_2(\omega) = \frac{2\sqrt{\gamma_1\gamma_2}}{\gamma_1 + \gamma_2 - 2i\omega},$$

$$g_2(\omega) = \frac{4\epsilon\sqrt{\gamma_1\gamma_2}}{(\gamma_1 + \gamma_2 - 2i\omega)^2}.$$

Here, ϵ is the single-pass parametric gain amplitude, proportional to the nonlinearity of the crystal and the pump power. ω_0 is the degenerate frequency of signal and idler. $\gamma_{1,2}$ are decay constants for modes $\hat{a}_{in}$ and $\hat{b}_{in}$ due to losses in the cavity system, with $\hat{b}_{in}$ being the operator for the unwanted vacuum mode coupled to the loss.

Using Eq. 2.61 we can calculate the overall signal enhancement as

$$R_{OPO} = \langle \hat{a}^\dagger_{out}(\omega_0 + \omega)\hat{a}_{out}(\omega_0 + \omega)\rangle = \frac{|r|^2\mathcal{F}^2}{\pi\, t_{rt}\mathcal{F}_0}, \qquad (2.62)$$

where $r \equiv \epsilon t_{rt}$ is the round-trip gain parameter with t_{rt} the round-trip time, $\mathcal{F} \equiv 2\pi/(\gamma_1 + \gamma_2)t_{rt}$ the finesse of the cavity and $\mathcal{F}_0 = \mathcal{F}(\gamma_2 = 0)$, the finesse without loss into the vacuum mode. Comparing this to the single pass rate without the cavity $R_{sp} = |r|^2\Omega/2\pi$, we find the average enhancement per mode is

$$E = \frac{R_{OPO}/\Delta\nu}{R_{sp}/\Omega} = \frac{\mathcal{F}^3}{2\mathcal{F}_0} \approx \mathcal{F}^2\gamma_2/\pi. \qquad (2.63)$$

Therefore the brightness will increase with the square of the finesse or in other words the square of the average number of bounces of the photon inside the cavity before leaving. We can intuitively understand this: if the amplitudes of all round-trips add up constructively (at resonance), the field inside the cavity increases linearly with the number of bounces and the down-converted field should increase quadratically. The loss of the system into the vacuum mode decreases the enhancement by a factor of $\mathcal{F}/\mathcal{F}_0$.

2.4 Single Photon Source Metrics

Single photons have become an important resource in quantum information and quantum communications in the past decade. Applications like linear optical quantum computing [52], quantum cryptography [53], quantum metrology [54] and quantum networks [55] are continuously driving the field towards high quality single photon sources. Trying to map the whole probability distribution of all possible states in order to characterise such a device is impractical. Alternatively, we can define certain figures of merit which need to fulfil standards dependent on the application, and evaluate systems by these accepted metrics. The upcoming pages will briefly introduce the photon as a Fock state of the electromagnetic field and then derive the formulas for four important characteristics of single photons for quantum networks: the linewidth, the spectral brightness, the multi-photon suppression and the indistinguishability.

The electromagnetic field in free space can be quantised by regarding the field vector $\mathbf{A}(\mathbf{r})$ as a Hilbert-space operator $\hat{\mathbf{A}}(\mathbf{r})$. This field operator can be written in terms of monochromatic modes of frequency ω_i as [56]

$$\hat{\mathbf{A}}(\mathbf{r}) = \sum_i \mathbf{A}_i(\mathbf{r})\hat{a}_i + \text{H.c.}, \tag{2.64}$$

where $\mathbf{A}_i(\mathbf{r})$ are the classical orthonormal mode functions that solve the Helmholtz equation

$$\left(\nabla^2 + \left(\frac{\omega_i}{c}\right)^2\right)\mathbf{A}_i(\mathbf{r}) = 0. \tag{2.65}$$

Further quantisation requires the classical Poisson brackets to be replaced by the quantum mechanical equal-time commutators. The bosonic commutation relations translate into $[\hat{a}_i, \hat{a}_{i'}^\dagger] = \delta_{i,i'}$, with $\hat{a}_i^\dagger$ and $\hat{a}_i$ the creation and annihilation operators of photons, respectively. The solution of Eq. 2.65 are plane waves with a wavevector $\mathbf{k}_i = \omega_i/c\, \mathbf{e}_i$, with $\mathbf{e}_i$ a vector pointing into the direction of propagation. For two orthogonal polarisations σ the field operator now becomes:

$$\hat{\mathbf{A}}(\mathbf{r}) = \sum_{\sigma=1}^{2} \int \frac{\mathrm{d}^3 k}{(2\pi)^{3/2}} \left(\frac{\hbar}{2\epsilon_0 kc}\right)^{1/2} \mathbf{e}_\sigma \hat{a}_\sigma(\mathbf{k}) e^{i\mathbf{k}\mathbf{r}} + \text{H.c.}, \tag{2.66}$$

with

$$[\hat{a}_\sigma(\mathbf{k}), \hat{a}_{\sigma'}^\dagger(\mathbf{k}')] = \delta_{i,i'}\delta(\mathbf{k} - \mathbf{k}'). \tag{2.67}$$

The Hamiltonian of the system describes its electromagnetic energy. It can be written as the sum over all harmonic oscillator Hamiltonians,

$$\hat{\mathbf{H}} = \sum_i \mathbf{H}_i = \sum_i \hbar\omega_i \left(\hat{a}_i^\dagger \hat{a}_i + \frac{1}{2}\right), \tag{2.68}$$

We can see that the number operator $\hat{n}_i = \hat{a}_i^\dagger \hat{a}_i$ commutes with the Hamiltonian and is therefore an observable of the system. Its eigenstates are called photon number or Fock states and satisfy $\hat{n}_i|n_i\rangle = n_i|n_i\rangle$, where $n_i \in \mathbb{N}_0$. The corresponding eigenvalues are

$$E_i = \hbar\omega_i \left(n_i + \frac{1}{2}\right). \tag{2.69}$$

It can be shown that $\hat{a}_i^\dagger |n_i\rangle$ is also an eigenstate of the harmonic oscillator Hamiltonian with Energy $E_i + \hbar\omega_i$. The energy levels thus form an equidistant ladder starting at the vacuum energy E_0 with spacing $\hbar\omega_i$, the photon energy.

A very important property of Fock states is the vanishing photon-number variance

$$\langle n_i| \left(\Delta\hat{n}_i\right)^2 |n_i\rangle = 0. \tag{2.70}$$

This means that, in contrast to classical light, quantised light in a Fock state has no fluctuations in photon number and subsequently a true single photon source has to produce Fock states.

2.4.1 Spectral Brightness

In order to achieve high repetition rates of experiments in quantum information or fast quantum communication, the brightness of a photon source is crucial. There exists a variety of definitions for the (spectral) brightness dependent on the application, even within the field of single photons. For example, in quantum dot emitters the brightness at the first lens is of particular interest. In narrow-band sources we are more interested in the actual amount of single photon pairs coming out of an optical fibre that can be used for further experiments. For all purposes throughout this thesis, the spectral brightness of a single photon source is defined as the number of photon pairs detected each second per optical bandwidth (in MHz) of one particular frequency mode and per mW of pump power,

$$[B] = \frac{\text{photon pairs/s}}{\text{mW} * \text{MHz}}, \tag{2.71}$$

within a symmetric coincidence window of twice the photon coherence time. The
bandwidth of SPDC of > 100 GHz and high pump powers usually lead to a spectral
brightness far below one. This is a serious limitation for the repetition rate of experiments when combining photon sources with quantum memories, where linewidths
are on the order of MHz. However, in cavity-enhanced SPDC the linewidth of the
single photons can be narrowed below the MHz level, depending on the application. Additionally, using a triply resonant cavity allows for constant creation of
pairs (100% duty cycle) and a much larger interaction region with the nonlinear crystal, significantly lowering the necessary pump power and increasing the
pair creation rate. With these features, our source is the first to achieve a spectral
brightness $B \sim 4000$ $\frac{\text{photon pairs / s}}{\text{mW} * \text{MHz}}$, as presented in Sect. 4.1.2. This work improved
the spectral brightness by a factor of ~ 3 compared to the brightest source so far
$(1410$ $\frac{\text{photon pairs / s}}{\text{mW} * \text{MHz}})$ [57] and even without corrections for the detector efficiency, we
were able to achieve $B = 2330$ $\frac{\text{photon pairs / s}}{\text{mW} * \text{MHz}}$.

2.4.2   Intensity Cross-Correlation Function $G_{s,i}^{(2)}(\tau)$: Linewidth

The temporal signal-idler intensity cross-correlation function $G_{s,i}^{(2)}(\tau)$ is of particular
interest since it allows easy access to important experimental parameters like the
FSR or the cavity decay rate of the biphoton state. It is measure by looking at
the coincidence rate for detecting a signal photon at time t and an idler photon at
time $t + \tau$. To calculate this temporal correlation function, we first need to derive
the biphoton wave function and the field operators for resonant signal and idler
modes [36, 58, 59].

Analogous to Eq. 2.66, but now assuming a linearly polarised wave with cross
section A travelling in the z direction, we can write the electric field operators for
signal and idler as

$$\hat{E}_{S,I}(z, t) = \left(\frac{\hbar \omega_{S,I}}{2\epsilon_0 cA} \right)^{1/2} \int_{-\infty}^{\infty} \frac{d\omega}{\sqrt{2\pi}} \hat{a}_{S,I} \left(\omega_{S,I} + \omega \right) e^{i(\omega_{S,I} + \omega)(z/c - t)}, \qquad (2.72)$$

with the corresponding central frequency $\omega_{S,I}$ and the operators $\hat{a}_{S,I}(\omega)$ following the
standard commutation relations (Eq. 2.67) for orthogonal polarisation of the signal
and idler photons in type II SPDC.

The biphoton wave function can be calculated by time evolution of the interaction Hamiltonian describing the down-conversion process. The general form of the
Hamiltonian for a crystal of susceptibility χ and length l is

$$H_{int} = \frac{\chi}{2l} \int_{-l}^{0} dz \left(\hat{E}_p^{cr} \hat{E}_S^{cr\dagger} \hat{E}_I^{cr\dagger} + H.c. \right), \qquad (2.73)$$

where $\hat{E}^{cr}$ is the field operator inside the crystal. Incorporating the explicit expressions for these field operators with energy and phase matching conditions into the Hamiltonian in the limit of an empty resonator before each emission event, we arrive at the final expression of the biphoton wave function [59]

$$
\begin{aligned}
|\Psi\rangle &= \frac{1}{i\hbar} \int_0^t dt' H_{int}(t')|0\rangle \\
&= \sum_{m_S} \sum_{m_I} \int_{-\infty}^{\infty} d\omega \int_{-\infty}^{\infty} d\omega' \frac{\pi\alpha\sqrt{\gamma_S\gamma_I} F_{m_S,m_I}(\omega,\omega')}{(\gamma_S/2 - i\omega)(\gamma_I/2 - i\omega')} \\
&\quad \times \delta\left(m_S\,\omega_{FSR,S} + \omega + m_I\,\omega_{FSR,I} + \omega'\right) \times \hat{a}_S^\dagger\left(\omega_{S,m_S} + \omega\right)\hat{a}_I^\dagger\left(\omega_{I,m_I} + \omega'\right)|0\rangle. \quad (2.74)
\end{aligned}
$$

Here, $m_{S,I} \in [-{}^{m_0}/_2; {}^{m_0}/_2]$ is the summation index over the number of cavity frequency modes m_0 within the phase matching bandwidth of the crystal. The constant α accounts for the strength of the pump field and the nonlinear interaction strength at signal and idler frequencies. $F_{m_S,m_I}(\omega,\omega')$ is a function that evaluates the collinear phase matching condition for type II SPDC inside the cavity. The damping rates and the FSR of the cavity are given as $\gamma_{S,I}$ and $\omega_{FSR,S}$ and $\omega_{FSR,I}$, respectively. In the regime far below threshold, we can assume the photon pair production rate to be much smaller than the cavity decay, leading to an empty cavity before every emission event. Subsequently, we can use perturbation theory for the evolution of an initial product state $|0\rangle(t=0) = |0\rangle_S \otimes |0\rangle_I$ into $|\Psi\rangle$. Equation 2.74 is very powerful and can be used to calculate a variety of properties of the single photon pair.

Combining Eqs. 2.72 and 2.74 we can now calculate a explicit expression of the temporal correlation function [60]

$$
G_{s,i}^{(2)}(\tau) = \langle\Psi|\hat{E}_I^\dagger(t)\hat{E}_S^\dagger(t+\tau)\hat{E}_S(t+\tau)\hat{E}_I(t)|\Psi\rangle \tag{2.75}
$$

for the specific case of a triply resonant OPO far below threshold in terms of frequency modes [36]:

$$
\begin{aligned}
G_{s,i}^{(2)}(\tau) \propto \Bigg| &\sum_{m_s,m_I} \frac{\sqrt{\gamma_S\gamma_I\omega_S\omega_I}}{\Gamma_s\Gamma_I} \\
&\times \begin{cases} e^{-2\pi\Gamma_s(\tau-(\tau_0/2))}\mathrm{sinc}(i\pi\tau_0\Gamma_s) & \forall \tau \geq \frac{\tau_0}{2} \\ e^{+2\pi\Gamma_i(\tau-(\tau_0/2))}\mathrm{sinc}(i\pi\tau_0\Gamma_i) & \forall \tau < \frac{\tau_0}{2} \end{cases} \Bigg|^2,
\end{aligned} \tag{2.76}
$$

where τ_0 corresponds to the temporal width of the peaks, accounting for the propagation delay between signal and idler in the crystal and $\forall k \in \{S, I\}$, $\Gamma_k = \frac{\gamma_k}{2} + i\,m_k\omega_{FSR,k}$. In terms of temporal modes we get [58]:

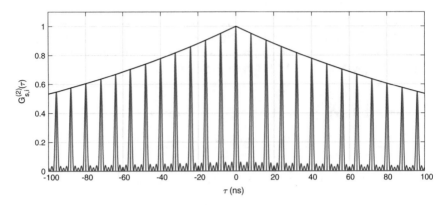

Fig. 2.15 Normalised second order cross-correlation function for parameters similar to our setup: $\gamma_S = \gamma_I = 1$ MHz, $t_{rt,S} = t_{rt,I} = 8$ ns and $\tau_0 = 3.7$ ps. The sum in Eq. 2.76 is run over one (black), three (blue) and 300 (red) frequency mode(s). We can clearly see that more frequency modes result in a narrowing of the temporal peaks around integer multiples of the free spectral range

$$
G_{s,i}^{(2)}(\tau) \propto
\begin{cases}
\displaystyle\sum_{j=1}^{\infty} \exp\left[-j\gamma_S t_{rt,S} - \frac{4(jt_{rt,S} - \tau)}{\Delta T^2} \right] & \forall \tau \geq \frac{\tau_0}{2} \\[4mm]
\displaystyle\sum_{j=-\infty}^{0} \exp\left[-j\gamma_I t_{rt,I} - \frac{4(jt_{rt,I} - \tau)}{\Delta T^2} \right] & \forall \tau < \frac{\tau_0}{2}
\end{cases}
\qquad (2.77)
$$

where ΔT characterises the effective resolution of the detectors.

In both cases, the signal-idler correlation function consists of equally spaced peaks of width τ_0 and spacing $t_{rt} = {}^{2\pi}/_{\omega_{FSR}}$, the round-trip time of the cavity. The peak height decays exponentially with the cavity damping rates $\gamma_{S,I}$ for positive and negative τ, respectively. Figure 2.15 depicts the temporal idler-signal intensity cross-correlation function for typical experimental parameters in our setup and compares the shape for many, a few and only one frequency mode. We can interpret the narrowing of the temporal peaks in the $G_{s,i}^{(2)}(\tau)$ function for a higher number of frequency modes as having more information on when exactly the photons left the cavity. For single mode operation, this information is completely lost and only the exponential decay remains. Therefore, the number of frequency modes contained in each single photon pair can be approximated by the peak width τ_0 in $G_{s,i}^{(2)}(\tau)$.

In an actual experiment, one needs to account for possible offsets, limited time resolution and jitter of the detectors as well as corrections for background (accidental) coincidences. These environmental factors slightly change $G_{s,i}^{(2)}(\tau)$, but the general shape stays the same. More on the experimental measurements and extraction of the crucial parameters of the cross-correlation function can be found in Sect. 4.1.1.

2.4.3 Intensity Auto-Correlation Function $g_{s,s}^{(2)}(\tau)$: Multi-photon Suppression

The individual photon statistics can be characterised by measuring the temporal idler-triggered intensity auto-correlation function $g_{s,s}^{(2)}(\tau)$. Its value at zero time delay, $g_{s,s}^{(2)}(0)$, is part of a convenient measure for the multi-photon suppression $(1 - g_{s,s}^{(2)}(0))$ of single photon sources, giving an estimate of unwanted higher order contributions towards the biphoton state. The auto-correlation function for photons can be measured in a relatively simple Hanbury Brown and Twiss experiment as shown in Fig. 2.16. The detection of an idler photon is used to trigger detectors in the signal arm which is evenly split up in two parts on a 50/50 beam splitter. Furthermore, one of the signal arms is delayed by a time τ. $g_{s,s}^{(2)}(\tau)$ now describes coincidences between an idler and a first signal photon arriving at time t and a second signal photon arriving at time $t + \tau$. It is given as [56, 61]

$$g_{s,s}^{(2)}(t + \tau | t) = \frac{\langle \hat{E}_S^\dagger(t) \hat{E}_S^\dagger(t + \tau) \hat{E}_S(t + \tau) \hat{E}_S(t) \rangle_{pm}}{\langle \hat{E}_S^\dagger(t) \hat{E}_S(t) \rangle_{pm} \langle \hat{E}_S^\dagger(t + \tau) \hat{E}_S(t + \tau) \rangle_{pm}}, \qquad (2.78)$$

where $\langle \bullet \rangle_{pm}$ is the average over the post-measurement state, only necessary for quantum states of light. In the regime of classical light sources, the auto-correlation function can only take values $g_{s,s}^{(2)}(\tau) \geq 1$. On the other hand, single photon Fock states allow for all positive values $g_{s,s}^{(2)}(\tau) \geq 0$, giving rise to an exclusive quantum region $1 > g_{s,s}^{(2)}(\tau) \geq 0$ where a pure single photon source has $g_{s,s}^{(2)}(0) = 0$. A variety of $g_{s,s}^{(2)}(0)$ values for different input states coming from a light bulb (thermal), laser (coherent) and single photon source (Fock) is given in Table 2.2. We can interpret the values as follows: thermal photons come in groups, an effect called bunching, whereas photons from a laser arrive randomly spaced in time and sometimes coincide. Single photons never coincide, i.e. the source ideally never emits more then one photon, which is called anti-bunching.

The following derivation of $g_{s,s}^{(2)}(\tau)$ for single photon sources is following the approach given in [61, 62]. The biphoton state of SPDC is a zero-mean Gaussian state whose only nonzero second-order moments are given as [63]:

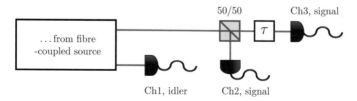

Fig. 2.16 Schematic of a Hanbury Brown and Twiss set-up. The intensity auto-correlation function with a delay τ is measured by looking at double and triple coincidences between the three channels

Table 2.2 Intensity
auto-correlation function

State	$g_{s,s}^{(2)}(0)$
Thermal	2
Coherent $|\alpha\rangle$	1
Single photon fock $|n = 1\rangle$	0

$$\langle \hat{E}_k^\dagger(t+\tau)\hat{E}_k(t)\rangle \equiv R(\tau)e^{i\omega_p\tau/2}, \qquad\qquad k = S, I, \qquad (2.79)$$

$$\langle \hat{E}_S^\dagger(t+\tau)\hat{E}_I(t)\rangle \equiv C(\tau)e^{-i\omega_p(t+\tau)/2}, \qquad\qquad\qquad (2.80)$$

representing the auto-correlation and the phase-sensitive cross-correlation function, respectively. In SPDC, we are interested in the low gain regime where the expressions for $R(\tau)$ and $C(\tau)$ can be approximated by:

$$R(\tau) = \begin{cases} R_0(1 + \tau/\Delta t) & -\Delta t < \tau \leq 0 \\ R_0(1 - \tau/\Delta t) & 0 < \tau \leq \Delta t \\ 0 & \text{else} \end{cases}, \qquad (2.81)$$

with R_0 the photon generation rate for signal and idler and $1/\Delta t$ the bandwidth of the single photons, and

$$C(\tau) = \begin{cases} \sqrt{R_0/\Delta t} & |\tau| < \Delta t/2 \\ 0 & \text{else} \end{cases}. \qquad (2.82)$$

It can be shown via a heuristic approach [64] or a discrete-mode formalism [62] that the post-measurement averaging for any operator $\hat{X}$ is given by

$$\langle \hat{X} \rangle_{pm} = \langle \hat{E}_i^\dagger(t_i)\hat{X}\hat{E}_i(t_i)\rangle / \sqrt{\langle \hat{E}_i^\dagger(t_i)\hat{E}_i(t_i)\rangle}. \qquad (2.83)$$

Without loss of generality, the idler-triggered intensity auto-correlation function (Eq. 2.78) for $t = 0$ can then be written as follows

$$g_{s,s}^{(2)}(\tau) = \frac{P_{ssi}(\tau)R(0)}{P_{si}(0)P_{si}(\tau)}, \qquad (2.84)$$

with $P_{si}(\tau)$ and $P_{ssi}(\tau)$ representing the double (signal-idler) and triple (signal-signal-idler) coincidence rate, respectively, for infinitely small coincidence windows and ideal detectors without jitter. We can use the quantum version of the Gaussian moment-factoring theorem [65] to obtain

$$P_{si}(\tau) = R_0^2 + C^2(\tau), \qquad (2.85)$$

$$P_{ssi}(\tau) = 2C(0)C(\tau)R(\tau) + R_0 \left[R_0^2 + R^2(\tau) + C^2(0) + C^2(\tau) \right]. \qquad (2.86)$$

These expressions cannot be simplified further, but their overall behaviour can be understood quite easily by looking at large and small time delays. For delays larger than the Δt, all non-constant terms become zero and the auto-correlation function $g_{s,s}^{(2)}(\tau > \Delta t) = 1$, independent of the creation rate. This implies that the coherence time of the photons is on the order of Δt. For small time delays around zero, the coherence function becomes [61]

$$g_{s,s}^{(2)}(0) = 2 - \frac{2C^4(0)}{\left[R^2(0) + C^2(0)\right]^2}. \tag{2.87}$$

It is clear that if $R^2(0) \ll C^2(0)$, or in other words $R_0 \ll {}^1/_{\Delta t}$, then $g_{s,s}^{(2)}(0) \approx 0$. This result is intuitively expected as there should not be any triple coincidence events for an ideal photon pair source.

In an actual experiment it is not possible to measure $P_{si}(\tau)$ or $P_{ssi}(\tau)$ directly, however, it is possible to measure time-averaged rates, which can be approximated as $N_{si}(\tau)$ or $N_{ssi}(\tau)$. Here, the detections of photons at certain times is extended to time intervals $[\tau - \tau_c, \tau + \tau_c]$ with a coincidence window width of $2\tau_c$. Additionally, we need to take the detector jitter into account, limiting the effective resolution to an interval of width $2\tau_d$ symmetrically around τ. The resulting functions are convolutions of $P_{si}(\tau)$ ($P_{ssi}(\tau)$) with two (three) detector response functions and one (two) coincidence window(s). We can then write the observed functions as

$$N_{si}(\tau) \approx \frac{1}{2\tau_c} \int_{\tau - \tau_c}^{\tau + \tau_c} d\tau' \bar{P}_{si}(\tau'), \tag{2.88}$$

$$N_{ssi}(\tau) \approx \frac{1}{(2\tau_c)^2} \int_{-\tau_c}^{\tau_c} dt_1 \int_{\tau - \tau_c}^{\tau + \tau_c} dt_2 \bar{P}_{ssi}(\tau), \tag{2.89}$$

where

$$\bar{P}_{si}(\tau) = \int dt_i \int dt_s u(t_i) u(t_s - \tau) P_{si}(\tau), \tag{2.90}$$

$$\bar{P}_{ssi}(\tau) = \int dt_i \int dt_{s_1} \int dt_{s_2} u(t_I) u(t_{s_1} - \tau) u(t_{s_2} - \tau) P_{ssi}(\tau). \tag{2.91}$$

Here, $u(t) = 1/(2\tau_d)$ if $|t| \leq \tau_d$, and zero otherwise. The times t_i, t_s correspond to detection times of idler and signal. $\bar{P}_{si}(\tau)$ and $\bar{P}_{ssi}(\tau)$ are the double and triple coincidence rates for detecting an idler photon at time zero and a signal at τ or signal and idler at time zero and additional signal at τ, respectively. The time averaged auto-correlation function is now:

$$g_{s,s}^{(2)}(\tau) = \frac{N_{ssi}(\tau)R_0}{N_{si}(0)N_{si}(\tau)}. \tag{2.92}$$

We can see from Eqs. 2.88 and 2.89, that the results for $g_{s,s}^{(2)}(\tau)$ will strongly depend on the chosen coincidence window τ_c. After some further algebra to calculate the convolutions we find [66]:

$$g_{s,s}^{(2)}(\tau) = \begin{cases} \sim \dfrac{1+(R_0\tau_c)^{-1}}{\left(1+(2R_0\tau_c)^{-1}\right)^2} & |\tau| < \tau_c - \delta t \\ 1 & |\tau| > \tau_c + \delta t \end{cases}, \qquad (2.93)$$

with $2\delta t$, the width of a transition region between short and long delays. There is no analytical solution for the region in between but we can see from Eq. 2.93 that the figure of merit, $g_{s,s}^{(2)}(0)$, increases linearly with the coincidence window for a constant photon detection rate in case of $R_0\tau_c \ll 1$, always fulfilled in our experiments. In reality, we can never reach $g_{s,s}^{(2)}(0) = 0$ due to the finite resolution of the detectors.

2.4.4   Hong-Ou-Mandel Interference: Indistinguishability

The Hong-Ou-Mandel (HOM) effect is a quantum interference effect that occurs when two indistinguishable photons overlap on a beam splitter. It was first observed in 1987 by Hong et al. [67] as a drop in the photon coincidence rate behind a beam splitter dependent on temporal delay between the two input ports. The indistinguishability of photons is of great importance in many applications like quantum information, computation or repeaters as it is the main ingredient of many protocols in those areas.

To model the interference we consider two photons, each in one spatial input mode a and b of a beam splitter. We can write such a state as

$$|\Psi_{in}\rangle = \hat{a}_i^\dagger \hat{b}_j^\dagger |0\rangle = |1_a; i\rangle |1_b; j\rangle = |1_a, 1_b\rangle_{ij} \qquad (2.94)$$

where the indices i, j are different values of the same arbitrary property of the photons, making them distinguishable. This could be either the temporal mode (e.g. arrival time at the beam splitter), the polarisation (e.g. horizontal or vertical), frequency mode or any other degree of freedom. An ideal lossless beam splitter with reflectivity η will now unitarily transform the creation operators as follows:

$$\hat{a}_i^\dagger \rightarrow i\sqrt{\eta}\hat{a}_i^\dagger + \sqrt{1-\eta}\hat{b}_i^\dagger, \qquad (2.95)$$

$$\hat{b}_j^\dagger \rightarrow \sqrt{1-\eta}\hat{a}_j^\dagger + i\sqrt{\eta}\hat{b}_j^\dagger, \qquad (2.96)$$

with the factor i coming for the phase acquired when reflecting light of a surface. In the special case of $\eta = 50\%$, the output state after the balanced beam splitter is

$$\begin{aligned} |\Psi_{out}\rangle &= \frac{1}{\sqrt{2}}\left(i\hat{a}_i^\dagger + \hat{b}_i^\dagger\right)\frac{1}{\sqrt{2}}\left(\hat{a}_j^\dagger + i\hat{b}_j^\dagger\right)|0\rangle \\ &= \frac{1}{2}\left(i\hat{a}_i^\dagger\hat{a}_j^\dagger + \hat{a}_j^\dagger\hat{b}_i^\dagger - \hat{a}_i^\dagger\hat{b}_j^\dagger + i\hat{b}_i^\dagger\hat{b}_j^\dagger\right)|0\rangle. \end{aligned} \qquad (2.97)$$

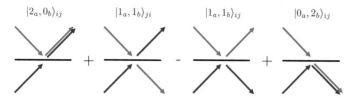

$$|2_a, 0_b\rangle_{ij} \qquad\qquad |1_a, 1_b\rangle_{ji} \qquad\qquad |1_a, 1_b\rangle_{ij} \qquad\qquad |0_a, 2_b\rangle_{ij}$$

Fig. 2.17 Four different outcomes of two photons entering a beam splitter (black horizontal line) from two input ports. The different colours indicate an arbitrary property that makes the two photons distinguishable. This could be frequency, but also the temporal mode, polarisation or any other degree of freedom. The + and − indicates the sign in front of the corresponding term in Eq. 2.97

This result leads to the expected four possible outcomes of this experiment as shown in Fig. 2.17.

Let us now consider the two scenarios of completely distinguishable photons, $i \neq j$, and perfectly indistinguishable photons, $i = j$, and look at the outcomes of Eq. 2.97. In the first case, the output state is

$$|\Psi_{dis}\rangle = \frac{1}{2}\left(i|2_a, 0_b\rangle_{ij} + |1_a, 1_b\rangle_{ji} - |1_a, 1_b\rangle_{ij} + i|0_a, 2_b\rangle_{ij}\right). \qquad (2.98)$$

In the HOM experiment, we are interested in the coincidence probability P_C with respect to the case without a beam splitter, P_0. We can see from Eq. 2.98, that for distinguishable photons only the middle terms contribute to the coincidence rate and hence $P_C = {}^{P_0}/2$, the probability decreases to 50% compared to an experiment without the beam splitter. In the case where both photons are indistinguishable, the middle terms cancel each other out via non-classical interference and the output is a two photon NOON state [68]

$$|\Psi_{indis}\rangle = \frac{i}{2}\left(|2_a, 0_b\rangle_{ii} + |0_a, 2_b\rangle_{ii}\right). \qquad (2.99)$$

Here, $P_C = 0$ as both photons always go to the same detector, because the amplitudes for both photons transmitted and both photons reflected cancel each other out.

So far, we only considered the two extreme cases of completely distinguishable and indistinguishable photons. However, it is possible to have partial indistinguishability in the desired degree of freedom. For example, partial distinguishability in polarisation can be achieved by changing one photon state from $|H\rangle \rightarrow \sqrt{\epsilon}\,|H\rangle + \sqrt{1-\epsilon}\,|V\rangle$, where $0 < \epsilon < 1$. In experiments, it is more common to change the arrival time of a photon at the beam splitter by introducing a delay τ between the photons. Assuming the photons have a certain temporal width δt linked to their frequency bandwidth, it is possible to control τ accordingly and tune the degree of indistinguishability as shown in Fig. 2.18. The coincidence probability is between $0 < P_C(\tau) < {}^1/2$ for partially distinguishable photons, where the transition shape crucially depends on the photon input state, as shown in the following.

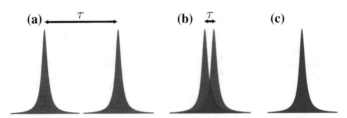

Fig. 2.18 Schematic of temporal distinguishability. Two photons of same width δt are delayed by τ. (**a**) For $\tau > \delta t$, the wave packets don't overlap and the photons are distinguishable. (**b**) In the region $\tau < \delta t$, the two photons are partially indistinguishable (blue region). (**c**) For perfect overlap, $\tau = 0$, the photons are maximally indistinguishable

To derive the shape of the HOM dip as a function of the delay time τ, we first define the electric field operators on both ports before (a and b) and after (1 and 2) the beam splitter as

$$\hat{E}_a(t) = \frac{1}{2\pi} \int d\omega \hat{a}(\omega) e^{-i\omega t}, \tag{2.100}$$

$$\hat{E}_b(t) = \frac{1}{2\pi} \int d\omega \hat{b}(\omega) e^{-i\omega t}, \tag{2.101}$$

and

$$\hat{E}_1(t) = \frac{1}{\sqrt{2}} \left[\hat{E}_a(t) + \hat{E}_b(t + \tau) \right], \tag{2.102}$$

$$\hat{E}_2(t) = \frac{1}{\sqrt{2}} \left[\hat{E}_a(t) - \hat{E}_b(t + \tau) \right], \tag{2.103}$$

where τ is the arrival time difference between the two photons. The coincidence rate can then be expressed as

$$R_C \propto \int dT \, G_{12}^{(2)}(t, t + T), \tag{2.104}$$

with the cross-correlation function

$$G_{12}^{(2)}(t, t + T) = \langle \Psi | \hat{E}_1^\dagger(t) \hat{E}_2^\dagger(t + T) \hat{E}_2(t + T) \hat{E}_1(t) | \Psi \rangle, \tag{2.105}$$

We can see that Eq. 2.105 is defined analogous to Eq. 2.75, but for different field operators $\hat{E}_1$ and $\hat{E}_2$. If we substitute Eqs. 2.100–2.103 and the state of spontaneous parametric down-conversion into Eq. 2.105 we obtain

$$G_{12}^{(2)}(t, t + T) = |g(T + \tau) - g(-T + \tau)|^2, \tag{2.106}$$

with

$$g(t) = \int d\omega \Phi(\omega)e^{i\omega t}. \tag{2.107}$$

Here, $\Phi(\omega)$ is the spectral amplitude function, describing the frequency spectrum of the single photon state. The coincidence rate now takes on its final version [69]

$$R_C(\tau) = \frac{1}{2} - \frac{1}{2} \frac{\mathrm{Re}\left[\int d\omega\ \Phi(-\omega)\Phi(\omega)e^{2i\omega\tau}\right]}{\left[\int d\omega\ |\Phi(\omega)|^2\right]}. \tag{2.108}$$

As pointed out before and proven by Eq. 2.108, the shape of the HOM dip for partially overlapping photons crucially depends on the frequency (temporal) shape of the photons. From here on we will focus on two important cases for $\Phi(\omega)$ in this thesis: the spectrum of SPDC and of a multimode OPO.

In the case of unfiltered SPDC, $\Phi(\omega)$ is given by the phase matching spectrum amplitude function of the form $\Phi(\omega) = \sqrt{\frac{\sigma}{\pi}}\mathrm{sinc}(\sigma\omega)$, with a full width at half maximum (FWHM) of the sinc^2 spectrum at $2.78/\sigma$. The corresponding coincidence rate is an inverse triangular function with a width 2σ at its base [70]:

$$R_{SPDC}(\tau) = \frac{1}{2} \times \begin{cases} |\tau|/\sigma & |\tau| \le \sigma \\ 1 & \text{else} \end{cases}. \tag{2.109}$$

The second important case for our experiment is the multi- (single-) mode output of an OPO. The theoretical predictions follow the calculations for mode-locked biphoton states [69, 71, 72], where $2N$ is the number of cavity frequency modes within the phase matching bandwidth with spacing $\omega_{FSR} = 2\pi\nu_{FSR}$ and the FWHM of each mode is $\Delta\omega = 2\pi\Delta\nu_{SP}$. For a frequency comb, as emitted by a multimode OPO, the spectrum amplitude function has the form:

$$\Phi(\omega) = \sum_{m=-N}^{N} h(\omega)f(\omega - m\omega_{FSR}) = \sum_{m=-N}^{N} \frac{\mathrm{sinc}(\sigma\omega)}{(\Delta\omega/2)^2 + (\omega - m\omega_{FSR})^2}, \tag{2.110}$$

where $h(\omega) = \mathrm{sinc}(\sigma\omega)$ is, again, the phase matching function of the SPDC and the cavity is modelled as Lorentzian peaks by $f(\omega) = \left[(\Delta\omega/2)^2 + (\omega - m\omega_{FSR})^2\right]^{-1}$. To obtain the rate $R_C(\tau)$ given in Eq. 2.108 we calculate both integrals:

$$\int d\omega\ |\Phi(\omega)|^2 = \frac{4\pi}{\Delta\omega^3} \sum_{m=-N}^{N} \mathrm{sinc}^2(m\sigma\omega_{FSR}), \tag{2.111}$$

$$\int d\omega \; \Phi(-\omega)\Phi(\omega)e^{2i\omega\tau} = \frac{4\pi e^{-\Delta\omega|\tau|}(1 + \Delta\omega|\tau|)}{\Delta\omega^3} \qquad (2.112)$$

$$\times \sum_{m=-N}^{N} \mathrm{sinc}^2(m\sigma\omega_{FSR})\cos(2m\omega_{FSR}\tau), \quad (2.113)$$

in order to obtain an explicit formula for the rate [69]:

$$R_C(\tau) = \frac{1}{2} - \frac{1}{2}\left\{\frac{e^{-\Delta\omega|\tau|}(1 + \Delta\omega|\tau|)}{\displaystyle\sum_{m=-N}^{N}\mathrm{sinc}^2(m\sigma\omega_{FSR})}\left[\sum_{m=-N}^{N}\mathrm{sinc}^2(m\sigma\omega_{FSR})\cos(2m\omega_{FSR}\tau)\right]\right\}.$$

$$(2.114)$$

Both cases, OPO and SPDC, are shown in Fig. 2.19 for typical experimental parameters. Equation 2.114 actually describes not only one single dip, but a pattern

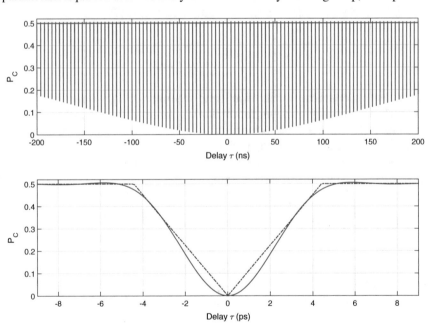

Fig. 2.19 HOM interference dips dependent on time delay for typical experimental parameters: $\sigma = 4.4$ ns, $\Delta\omega = 2\pi * 1$ MHz, $\omega_{FSR} = {}^{2\pi}/_8$ GHz, and $2N = 800$ modes. Top: HOM dip revivals with decreasing visibility (increasing maximum) dependent on the bandwidth of the individual modes as per Eq. 2.114. The spacing of the dips is $t_{rt}/2 = 4$ ns. Bottom: Comparison of the dip expected from the same type II SPDC crystal (blue, dashed) and the central dip of the OPO (red, solid). We can see that the cavity does not notably change the width of the individual dip. However, the shape gets altered significantly, especially around the minimum

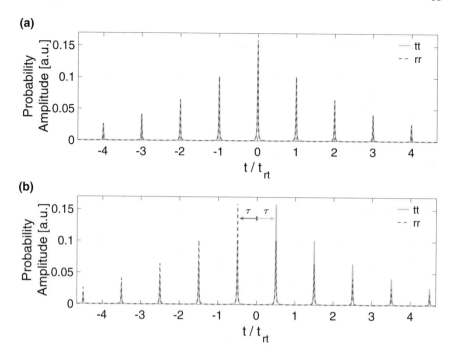

Fig. 2.20 Detection probability amplitudes of the reflected-reflected (rr) and transmitted-transmitted (tt) photon paths in the HOM experiment. (**a**) No temporal delay, $\tau = 0$. Amplitudes cancel each other out perfectly. (**b**) Revivals occur when the photons are delayed by an integer multiple of $t_{rt}/2$, $\tau = m^* t_{rt}/2$ with $m \in \mathbb{Z}$, here shown for $m = 1$. Both detection probability amplitudes shift by the implemented delay but in opposite directions, overlapping the combs again

of multiple dips with increasing minimum and a dip revival period of half the round-trip time of the cavity. Intuitively one would expect a revival period of a full round-trip time as photons that leave the cavity one round-trip earlier are interacting with photons that leave one round-trip later. The reason for this discrepancy is as follows: for simultaneously arriving photons, all temporal components of the biphoton state overlap as shown in Fig. 2.20(a), and thus interfere. For a temporal delay τ between the photons, the tt and rr detection probability amplitudes are both shifted, however, in opposite directions. This leads to a relative shift of 2τ so that the amplitudes re-overlap for delays matching an integer multiple of $t_{rt}/2$, $\tau = m^* t_{rt}/2$ with $m \in \mathbb{Z}$, as illustrated in Fig. 2.20(b), resulting in HOM dip revivals. The somehow counter intuitive revival period is ensured by the temporal entanglement of the two photons. The relative shift of the biphoton detection amplitudes lowers their quantitative overlap—especially visible around the centre of Fig. 2.20(b)—which further decreases with increasing temporal shift. This mismatch subsequently reduces the visibility of the interference, defined below.

In order to quantify the indistinguishability of two photons, we can define the visibility of the HOM dip as

$$V = \frac{R_{max} - R_{min}}{R_{max} + R_{min}},\tag{2.115}$$

with $R_{max} \equiv R_C(\infty)$ and $R_{min} \equiv R_C(0)$. In HOM interference experiments, the visibility is usually defined as $V = \frac{R_{max} - R_{min}}{R_{max}}$, but in the literature for narrowband SPDC the definition according to Eq. 2.115 is more common, e.g. [33]. Using the standard definition would further increase the obtained values presented in Sect. 4.2.2. The visibility gives us a good figure of merit to characterise and compare different photon sources from different architectures.

References

1. Drever, R., et al.: Laser phase and frequency stabilization using an optical resonator. Appl. Phys. B **31**, 97–105 (1983)
2. Siegman, A.: Lasers, 1st edn. University Science Books (1986)
3. Abbott, B.P., et al.: Observation of gravitational waves from a binary black hole merger. Phys. Rev. Lett. **116**, 061102 (2016)
4. Wheeler, M.D., Newman, S.M., Orr-Ewing, A.J., Ashfold, M.N.: Cavity ring-down spectroscopy. J. Chem. Soc. Faraday Trans. **94**, 337–351 (1998)
5. Saleh, B., Teich, M.: Fundamentals of Photonics (Wiley Series in Pure and Applied Optics), 2nd edn. Wiley (2007)
6. Abitan, H., Skettrup, T.: Laser resonators with several mirrors and lenses with the bow-tie laser resonator with compensation for astigmatism and thermal lens effects as an example. J. Opt. A Pure Appl. Opt. **7**, 7 (2005)
7. Barger, R.L., Sorem, M., Hall, J.: Frequency stabilization of a cw dye laser. Appl. Phys. Lett. **22**, 573–575 (1973)
8. Hansch, T., Couillaud, B.: Laser frequency stabilization by polarization spectroscopy of a reflecting reference cavity. Opt. Commun. **35**, 441–444 (1980)
9. Harvey, M.D., White, A.G.: Frequency locking by analysis of orthogonal modes. Opt. Commun. **221**, 163–171 (2003)
10. Black, E.: An introduction to pound-drever-hall laser frequency stabilization. Am. J. Phys. **69**, 79 (2001)
11. Fox, R.W., Oates, C., Hollberg, L.W.: Stabilizing diode lasers to high-finesse cavities. Exp. Methods Phys. Sci. **40**, 1 (2003)
12. Demtröder, W.: Laser spectroscopy: basic concepts and instrumentation. In: Advanced Texts in Physics, 5th edn. Springer, Berlin (2007)
13. Ilten, P.J.: Doppler-free Spectroscopy of Rubidium (2007)
14. Feynman, R., Leighton, R., Sands, M.: The Feynman Lectures on Physics, vol. 1, chap. 31. Addison-Wesley (1963)
15. Torii, Y., Tashiro, H., Ohtsubo, N., Aoki, T.: Laser-phase and frequency stabilization using atomic coherence. Phys. Rev. A **86**, 033805 (2012)
16. Goodwin, G.C., Graebe, S.F. Salgado, M.E.: Control System Design, 1st edn. Prentice Hall PTR, Upper Saddle River (2000)
17. Nagrath, I.: Control Systems Engineering. New Age International (P) Limited (2006)
18. Bechhoefer, J.: Feedback for physicists: a tutorial essay on control. Rev. Mod. Phys. **77**, 783–836 (2005)
19. Ziegler, J.G., Nichols, N.B.: Optimum settings for automatic controllers. Trans. ASME **64**, 759–768 (1942)
20. Menzel, R.: Photonics: linear and nonlinear interactions of laser light and matter. In: Advanced Texts in Physics, 2nd edn. Springer, Berlin (2007)

21. Boyd, R.W.: Nonlinear Optics, 3rd edn. Academic Press (2008)
22. Ding, Y.J., Kaplan, A.E.: Nonlinear magneto-optics of vacuum: second-harmonic generation. Phys. Rev. Lett. **63**, 2725–2728 (1989)
23. Boyd, G.D., Kleinman, D.A.: Parametric interaction of focused gaussian light beams. J. Appl. Phys. **39**, 3597–3639 (1968)
24. Targat, R.L., Zondy, J.-J., Lemonde, P.: 75%-efficiency blue generation from an intracavity ppKTP frequency doubler. Opt. Commun. **247**, 471–481 (2005)
25. Armstrong, J.A., Bloembergen, N., Ducuing, J., Pershan, P.S.: Interactions between light waves in a nonlinear dielectric. Phys. Rev. **127**, 1918–1939 (1962)
26. Lim, E.J., Fejer, M.M., Byer, R.L.: Second-harmonic generation of green light in periodically poled planar lithium niobate waveguide. Electron. Lett. **25**, 174–175 (1989)
27. Kato, K., Takaoka, E.: Sellmeier and thermo-optic dispersion formulas for KTP. Appl. Opt. **41**, 5040–5044 (2002)
28. Fedrizzi, A., et al.: Anti-symmetrization reveals hidden entanglement. New J. Phys. **11**, 103052 (2009)
29. Langford, N.K.: Encoding, manipulating and measuring quantum information in optics. Ph.D. thesis, The University of Queensland, Brisbane, Australia (2007)
30. Resch, K.J.: Making photons talk to each other: nonlinear optics in the quantum domain. Ph.D. thesis, University of Toronto, Canada (2003)
31. Ou, Z.Y., Lu, Y.J.: Cavity enhanced spontaneous parametric down-conversion for the prolongation of correlation time between conjugate photons. Phys. Rev. Lett. **83**, 2556–2559 (1999)
32. Kuklewicz, C.E., Wong, F.N.C., Shapiro, J.H.: Time-bin-modulated biphotons from cavity-enhanced down-conversion. Phys. Rev. Lett. **97**, 223601 (2006)
33. Wolfgramm, F., et al.: Bright filter-free source of indistinguishable photon pairs. Opt. Express **16**, 18145–18151 (2008)
34. Bao, X.-H., et al.: Generation of narrow-band polarization-entangled photon pairs for atomic quantum memories. Phys. Rev. Lett. **101**, 190501 (2008)
35. Yang, J., et al.: Experimental quantum teleportation and multiphoton entanglement via interfering narrowband photon sources. Phys. Rev. A **80**, 042321 (2009)
36. Scholz, M., Koch, L., Benson, O.: Statistics of narrow-band single photons for quantum memories generated by ultrabright cavity-enhanced parametric down-conversion. Phys. Rev. Lett. **102**, 063603 (2009)
37. Scholz, M., Koch, L., Ullmann, R., Benson, O.: Single-mode operation of a high-brightness narrow-band single-photon source. Appl. Phys. Lett. **94** (2009)
38. Haase, A., Piro, N., Eschner, J., Mitchell, M.W.: Tunable narrowband entangled photon pair source for resonant single-photon single-atom interaction. Opt. Lett. **34**, 55–57 (2009)
39. Wang, F.-Y., Shi, B.-S., Guo, G.-C.: Generation of narrow-band photon pairs for quantum memory. Opt. Commun. **283**, 2974–2977 (2010)
40. Zhang, H., et al.: Preparation and storage of frequency-uncorrelated entangled photons from cavity-enhanced spontaneous parametric downconversion. Nat. Photonics **5**, 628–632 (2011)
41. Wolfgramm, F., de Icaza Astiz, Y.A., Beduini, F.A., Cerè, A., Mitchell, M.W.: Atom-resonant heralded single photons by interaction-free measurement. Phys. Rev. Lett. **106**, 053602 (2011)
42. Fekete, J., Rieländer, D., Cristiani, M., de Riedmatten, H.: Ultranarrow-band photon-pair source compatible with solid state quantum memories and telecommunication networks. Phys. Rev. Lett. **110**, 220502 (2013)
43. Zhou, Z.-Y., Ding, D.-S., Li, Y., Wang, F.-Y., Shi, B.-S.: Cavity-enhanced bright photon pairs at telecom wavelengths with a triple-resonance configuration. J. Opt. Soc. Am. B **31**, 128–134 (2014)
44. Rambach, M., Nikolova, A., Weinhold, T.J., White, A.G.: Sub-megahertz linewidth single photon source. APL Photonics **1** (2016)
45. Fabre, C.: Classical and Quantum Aspects of C.W. Parametric Interaction in a Cavity, pp. 293–318. Springer Netherlands (1998)
46. Martinelli, M., Zhang, K.S., Coudreau, T., Maître, A., Fabre, C.: Ultra-low threshold CW triply resonant OPO in the near infrared using periodically poled lithium niobate. J. Opt. A Pure Appl. Opt. **3**, 300 (2001)

47. Träger, F.: Handbook of Lasers and Optics. In: Springer Handbooks, 1st edn. Springer, Berlin (2007)
48. Collett, M.J., Gardiner, C.W.: Squeezing of intracavity and traveling-wave light fields produced in parametric amplification. Phys. Rev. A **30**, 1386–1391 (1984)
49. Wu, L.-A., Kimble, H.J., Hall, J.L., Wu, H.: Generation of squeezed states by parametric down conversion. Phys. Rev. Lett. **57**, 2520–2523 (1986)
50. Breitenbach, G., Schiller, S., Mlynek, J.: Measurement of the quantum states of squeezed light. Nature **387**, 471–475 (1997)
51. Zhang, Y., et al.: Experimental generation of bright two-mode quadrature squeezed light from a narrow-band nondegenerate optical parametric amplifier. Phys. Rev. A **62**, 023813 (2000)
52. Knill, E., Laflamme, R., Milburn, G.J.: A scheme for efficient quantum computation with linear optics. Nature **409**, 46–52 (2001)
53. Gisin, N., Ribordy, G., Tittel, W., Zbinden, H.: Quantum cryptography. Rev. Mod. Phys. **74**, 145–195 (2002)
54. Giovannetti, V., Lloyd, S., Maccone, L.: Quantum metrology. Phys. Rev. Lett. **96**, 010401 (2006)
55. Duan, L.M., Lukin, M.D., Cirac, J.I., Zoller, P.: Long-distance quantum communication with atomic ensembles and linear optics. Nature **414**, 413–418 (2001)
56. Loudon, R.: The Quantum Theory of Light, 1st edn. Clarendon Press, Oxford (1973)
57. Chuu, C.-S., Yin, G.Y., Harris, S.E.: A miniature ultrabright source of temporally long, narrowband biphotons. Appl. Phys. Lett. **101**, 051108 (2012)
58. Herzog, U., Scholz, M., Benson, O.: Theory of biphoton generation in a single-resonant optical parametric oscillator far below threshold. Phys. Rev. A **77**, 023826 (2008)
59. Scholz, M., Koch, L., Benson, O.: Analytical treatment of spectral properties and signal-idler intensity correlations for a double-resonant optical parametric oscillator far below threshold. Opt. Commun. **282**, 3518–3523 (2009)
60. Glauber, R.J.: The quantum theory of optical coherence. Phys. Rev. **130**, 2529–2539 (1963)
61. Bocquillon, E., Couteau, C., Razavi, M., Laflamme, R., Weihs, G.: Coherence measures for heralded single-photon sources. Phys. Rev. A **79**, 035801 (2009)
62. Razavi, M., et al.: Characterizing heralded single-photon sources with imperfect measurement devices. J. Phys. B At. Mol. Opt. Phys. **42**, 114013 (2009)
63. Wong, F.N.C., Shapiro, J.H., Kim, T.: Efficient generation of polarization-entangled photons in a nonlinear crystal. Laser Phys. **16**, 1517–1524 (2006)
64. Nielsen, M.A., Chuang, I.L.: Quantum Computation and Quantum Information, 10th edn. Cambridge University Press, New York (2011)
65. Shapiro, J.H., Sun, K.-X.: Semiclassical versus quantum behavior in fourth-order interference. J. Opt. Soc. Am. B **11**, 1130–1141 (1994)
66. Bettelli, S.: Comment on "coherence measures for heralded single-photon sources". Phys. Rev. A **81**, 037801 (2010)
67. Hong, C.K., Ou, Z.Y., Mandel, L.: Measurement of subpicosecond time intervals between two photons by interference. Phys. Rev. Lett. **59**, 2044–2046 (1987)
68. Kok, P., Lee, H., Dowling, J.P.: Creation of large-photon-number path entanglement conditioned on photodetection. Phys. Rev. A **65**, 052104 (2002)
69. Xie, Z., et al.: Harnessing high-dimensional hyperentanglement through a biphoton frequency comb. Nat. Photonics **9**, 536–542 (2015)
70. Branczyk, A.: Non-classical states of light. Ph.D. thesis, University of Queensland (2010)
71. Shapiro, J.: Coincidence dips and revivals from a type-ii optical parametric amplifier. In: Nonlinear Optics: Materials, Fundamentals and Applications, FC7. Optical Society of America (2002)
72. Lu, Y.J., Campbell, R.L., Ou, Z.Y.: Mode-locked two-photon states. Phys. Rev. Lett. **91**, 163602 (2003)

Chapter 3
Design of a Narrowband Single Photon Source

Photons are the ideal carriers for quantum information: they travel at the speed of light, enabling the fastest communication possible, their interaction with the environment is weak, resulting in low decoherence, and they allow simple encoding of quantum information in multiple degrees of freedom, e.g. in polarisation or frequency. The development of optical fibres with losses <0.15 dB/km at telecom wavelengths made communication distances ~100 km [1–3] and most recently up to 300 km [4] possible, fundamentally limited by chromatic and modal dispersion, scattering and absorption [5]. This is insufficient for the implementation of large global networks and therefore the original information needs to be restored on a regular basis via so-called quantum repeaters [6–11]. The repeater can either be a series of high fidelity gates ($>99\%$) in an all optical approach [12], or the information is stored locally in a quantum memory, generally in some type of atomic transitions, purified, and then a new photon is emitted, carrying the initial information. To build such a network of quantum nodes, we have to achieve efficient interaction between atoms and single photons. This is not a trivial task because photons usually have bandwidths 5–6 orders of magnitude larger than the transitions they are aiming at. Cavity-enhanced SPDC can solve this problem while maintaining high rates of photon pair creation as discussed in Sect. 2.3.4. The upcoming sections will present the design considerations of the optical and the electronic control system in order to build a narrowband single photon source suitable for quantum memories based on the rubidium (Rb) D_1 transition at 795 nm.

3.1 Optical Parametric Oscillator Design

There are a couple of preliminary decisions to be made in order to build an OPO: which crystal is suitable for the wavelengths involved and what type of SPDC do we want? Which design should we use for the cavity: standing wave or ring cavity?

© Springer Nature Switzerland AG 2018
M. Rambach, *Narrowband Single Photons for Light-Matter Interfaces*,
Springer Theses, https://doi.org/10.1007/978-3-319-97154-4_3

How do we compensate for birefringence and dispersion inside the crystal? Doubly or triply resonant cavity? What linewidth are we aiming for and how can we obtain it? It is important to answer these questions in advance with regards to the application of the source. The following sections will be about the solutions we implemented in our experiment.

3.1.1   Crystal

The choice of the best suitable nonlinear crystal is crucial for efficient conversion of the pump light into single photons while simultaneously keeping the cavity losses low. Additionally, the material should introduce no spatial walk off between the signal and idler beam but allow for easy separation of the two and the phase matching temperature should be sufficiently over the dew point of Brisbane, to stop water condensation onto the crystal surfaces.

A crystal that fulfils the requirements is periodically poled potassium titanyl phosphate (ppKTP) grown in a geometry for type II collinear phase matched SPDC from 397.5 to 795 nm. The crystal has a high nonlinearity and low losses due to small absorption at the conversion wavelength. First order quasi-phase matching is achieve with a poling period $\Lambda_{QPM} = 8.8\,\mu m$ for SPDC. The FWHM phase matching bandwidth of our 25 mm long crystal is 100 GHz (see Sect. 2.3.2), very narrow for SPDC but still orders of magnitude above the atomic transition in rubidium [13]. Brute-force filtering to a sub-MHz level would reduce the spectral brightness by a factor $> 10^5$.

The collinear emission of the photons eliminates spatial walk-off between the optical fields involved, making angle tuning largely redundant. However, there will still be a walk-off in the direction of travel between the orthogonally polarised signal and idler photons due to birefringence, resulting in some degree of distinguishability. We can estimate the walk-off by looking at the Sellmeier equations for the ordinary and extraordinary refractive index and calculating the difference in travel time: for our crystals, the maximal delay is $\tau_{walk-off} = 7.4\,ps$, for a photon pair created at the very beginning of the crystal. This is several orders of magnitude lower than the expected coherence time of the photons, thus it can be neglected for the distinguishability measurements. Unfortunately, the effect is significant enough to affect the resonance condition of the cavity, especially as the delay accumulates with the number of round-trips. Different schemes to compensate for this birefringence walk-off are presented in Sect. 3.1.3. The problem could generally be avoided by using type I SPDC, but deterministic separation of frequency-degenerate signal and idler is only possible if orthogonally polarised photons (type II) are split after the leaving the cavity on a polarising beam splitter.

In order to determine the crystal temperature for optimal phase-matching, we measure the intensity of the inverse process, SHG. A laser beam at 795 nm with roughly the expected spot size and position calculated for the SPDC process, is coupled backwards into the cavity. To get a first idea of the approximate phase matching temperature, the SHG output is monitored on a camera while rapidly changing the crystal temperature. Although the SHG power output is fairly small, a sensitive CCD camera can detect the converted light. Next, we can tune the temperature to the measured maximum and align a sensitive photodiode to properly characterise the temperature dependent phase matching envelope. Due to space restrictions of the SPDC cavity in the experiment, the measurements are performed for the similar—easily accessible—SHG crystal in our setup. Nevertheless, the full temperature phase-matching range of 1.7 °C is the same for both crystals, with the centre temperature for optimal phase matching around 57.3 and 41.4 °C for SHG and SPDC, respectively, different mainly due to the individual approaches in the generation of the poling period but still far above the dew point. Further details on the control of the temperature can be found in Sect. 3.3.3 of this chapter.

3.1.2 Cavity Parameters

A well-considered cavity design is essential to the conversion efficiency and the linewidth of the photon pairs. As mentioned previously, the first choice that needs to be made is between standing wave and ring cavity. Both types have been demonstrated in the field of narrowband single photon sources, but we decided on using a ring resonator. Ring cavities offer multiple advantage compared to standing wave resonators: the losses per round-trip from the crystal and its surfaces are lower, as the light only travels once through the crystal per round trip. Moreover, there are extra input and output angles for coupling or for distinction between opposing directions of travel of the light because the mirror surfaces are slightly angled to each other. This is a big advantage as unwanted photons created by back reflections of the pump light are spatially separated, with demonstrated path isolations up to 50 dB [14].

There is a variety of ring cavity designs, as discussed in Sect. 2.1.1, with triangular and bow-tie structure probably the most prominent. The bow-tie design has several benefits with regards to mode matching, resonance condition and geometry of the cavity, all linked to a smaller angle between the incoming and outgoing light beam. A small angle of incident, present for at least one out of the three mirrors of a triangular cavity, is unfavourable when we try to achieve simultaneous resonance of different polarisations and wavelengths. Additionally, having two focuses in the space between the mirrors allows us to use one focus for the crystal and the second larger focus for mode-matching, enabling the implementation of lenses with long focal length and increasing the tolerance on the beam size in the coupling process. Lastly, the physical optical path length of a bow-tie cavity can be fairly long, while still keeping the overall design compact.

Figure 3.1 shows a schematic of the cavity-enhanced SPDC setup. The cavity is symmetrical and consists of four mirrors, two plane ($M_{1,2}$) and two curved with a radius of curvature of 200 mm ($M_{3,4}$). All mirrors are dual coated for the pump and single photon wavelengths with their reflectivities given in Table 3.1. For completeness of the cavity loss budget and neglecting the low absorption losses for now,

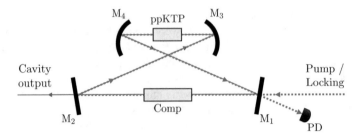

Fig. 3.1 Conceptual scheme of the triply resonant OPO setup. ppKTP, nonlinear crystal; Comp, birefringence compensating element(s); $M_{1,2,3,4}$, plane and curved cavity mirrors; PD, photodiode. The cavity is designed to let single photons and pump light escape through different mirrors, M_1 and M_2, respectively. Resonance of signal, idler and pump light is achieved by active stabilisation of the cavity length and crystal temperature, paired with careful crystal alignment

Table 3.1 OPO mirror reflectivities at 397.5 and 795 nm. $M_{1,2}$ plane, $M_{3,4}$ plano-concave mirrors

	Coating at 397.5 nm	Coating at 795 nm
M_1 (Incoupling)	$PR = 98.0\% \pm 0.4\%$	$HR > 99.9\%$
M_2 (Outcoupling)	$HR > 99.85\%$	$PR = 99.0\% \pm 0.2\%$
$M_{3,4}$	$HR > 99.85\%$	$HR > 99.9\%$
Crystal	$AR < 0.5\%$	$AR < 0.2\%$

the coatings of the crystal were added. The system is designed so that the single photons predominantly leave through the partially reflective outcoupling mirror M_2, with only a small portion of the pump field leaking out the same direction due to the high reflectivity of M_2 compared to the incoupling mirror M_1 at 397.5 nm.

The cavity length, curved mirror distance and radius are roughly chosen to optimise the conversion according to the Boyd-Kleinman criterion [15] for our crystal,

$$\omega_{0,opt} = \sqrt{\frac{l\lambda_0}{2\pi\, 2.84\, n}} = 24.6\,\mu m.$$ The cavity has a total length of $\sim$1219 mm, leading to an optimal beam waist for either 181 or 228 mm separation of the two curved mirrors. To avoid effects like thermal lensing and grey-tracking at very small beam waists with high power and in agreement with the finding that high efficiency conversion is still given up to twice the Boyd-Kleinman criterion [16], we chose a mirror spacing of 216 mm leading to a waist size of 37 μm. Figure 3.2a shows the dependence of the waist between the two curved mirrors as a function of their separation for a fixed cavity length. The possible positions to fulfil the Boyd-Kleinman criterion are marked in red and the actual spacing (216 mm) is highlighted in green. The waist size is fairly constant over a couple of mm around the chosen position, slightly lifting the strict requirements for positioning of the two curved mirrors.

The spatial beam profile of the cavity mode for the optimised distances between all mirrors is shown in Fig. 3.2b. As mentioned before in Sect. 2.1.2, the two curved mirrors of the cavity will effectively act as lenses with a focal length equal to half the radius of curvature. This means that we are expecting two focuses halfway between

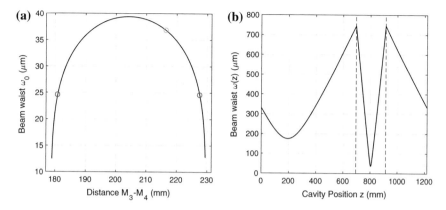

Fig. 3.2 Cavity length calculations. (**a**) Beam waist ω_0 as a function of the separation between the curved mirrors for a fixed cavity length of 1219 mm. The red dots indicate the spacing and waist according to the Boyd-Kleinman criterion, the green dot is the actual spacing chosen in the OPO: 216 mm. (**b**) Calculated beam profile as a function of distance $\omega(z)$ for the actual curved mirror spacing starting ($z = 0$ mm) and ending ($z = 1219$ mm) at the incoupling mirror M_1. We can see two focuses, one used for mode-matching of the incoming light around $z = 195$ mm and the second inside the SPDC crystal around $z = 804$ mm. The dashed vertical lines indicate the positions of the curved mirrors, effectively acting as lenses

the mirror pairs $M_{1,2}$ and $M_{3,4}$ for our symmetric cavity. The size of the larger waist between the two plane mirrors is calculated to be 175 μm, half an order of magnitude higher than the tight focus for the SPDC crystal. This simplifies the coupling drastically as a small mismatch of the spot size between cavity mode and incident beam still provides large overlap. Figure 3.2b also demonstrates the self-producing standing wave that is formed inside the cavity as $\omega(0\,\text{mm}) = \omega(1219\,\text{mm})$, the beam waist is of the same size and wavefront curvature after a physical round-trip.

Another important component of the OPO is indicated as a birefringence compensating element in Fig. 3.1. We have already found in Sect. 3.1.1 that the delay of signal and idler due to birefringence can be neglected for the distinguishability of narrowband photons. However, in order to stabilise the cavity, doubly or triply resonant, we need to compensate the effect. There are different approaches to solve this issue, found in all type II setups, using compensation crystals [17–19], angular and/or temperature tuning [20–22] or in our case a half-wave plate [23]. Details on these methods are discussed in Sect. 3.1.3.

Our bow-tie cavity is designed to operate at triple resonance. This requires far lower pump power compared to the double resonant case and has the major advantage that we do not need to divide the operational time of the source into two parts: a locking component, where the cavity length is stabilised and no photons are created, and a production component, where photon pairs are produced but the length of the cavity can change. Therefore, our source has 100% duty cycle, meaning that it can probabilistically create single photons at any given time. We ensure resonance with the pump light at 397.5 nm by adjusting the length of the cavity. This is

achieved through controlling the position of the curved mirror M_4 (Fig. 3.1) which
is mounted on a piezo-electric transducer. We derive the error signal for the control
loop from a photodiode combined with a Labview-controlled FPGA system via the
PDH method, introduced in Sect. 2.2.1. The control electronics and the experimental
implementation of the lock is described in detail in Sects. 3.3 and 3.4.2. The compensating element generally only overlaps the resonances of signal and idler photons,
but slight adjustments of the temperature combined with alignment optimisation of
the ppKTP can accomplish triple resonance.

All four mirrors, the holder for the crystal and the compensation element are
mounted on one block of Invar for increased stability. Invar is a nickel-iron alloy
with a linear, but more importantly, exceptionally low expansion over a wide range
of temperatures. The high density and low length change with temperature make
Invar the perfect material for a solid foundation of the cavity, cancelling out low
frequency noise and thermal drifts in the laboratory. Additionally, two boxes made
of acrylic are placed over the optical elements of the cavity and the experimental
breadboard, sitting on the edge of the Invar block and the optical table, respectively.
The effect of the boxes on the temperature stability of the crystal was measured for
three different scenarios: no box, only the small box around the cavity, and both
boxes in place. Table 3.2 shows the average actual temperature T_{av} and the standard
deviation σ_T of the measurements. More details on the effect of the boxes and the
general temperature stabilisation can be found in Sect. 3.3.3.

In order to roughly estimate a lower bound on the expected linewidth of the OPO,
we calculate the FSR and finesse of the cavity from its length and the specifications
of its elements, respectively. Following Table 3.1 and Eq. 2.3 we can find the cavity
gain parameter from Eq. 2.4 to be

$$g_{rt} = \sqrt{R_{M_1}^3 R_{M_2}(1 - R_{cr})^2 (1 - P_{abs})} > 0.99, \qquad (3.1)$$

with R, the reflectivities of the mirror or crystal surfaces and P_{abs} the absorption
inside the crystal, given in Ref. [24]. The result of Eq. 3.1 leads to a lower limit
for the finesse of the OPO of $\mathcal{F} = 323$. With a physical length of roughly 1219 mm
and neglecting birefringence effects for now, we get a FSR of $\nu_{FSR} = 244$ MHz.
The corresponding expected linewidth is $\Delta\nu = 755$ kHz at a wavelength of 795 nm,
well suited for integration with rubidium-based quantum memories demanding sub-
natural linewidths, such as gradient echo memories (GEM) [25, 26].

Table 3.2 Temperature control performance dependent on cover box for a set temperature of
41.4 °C

	No box	Box 1	Box 1 and 2
T_{av} (°C)	41.387	41.388	41.388
σ_T (mK)	8.2	1.5	<0.5

3.1.3 Birefringence Compensation

In the practical implementation of an OPO for type II SPDC, we find two challenges related to the frequency and the polarisation of the created photon pair. The first problem is dispersion: the speed of light inside an optical medium is dependent on the frequency. Fortunately, this effect is very small in ppKTP and our photons are very narrow in frequency, so we can neglect this issue. We already briefly discussed the second effect, birefringence, in Sect. 3.1.1 and found that the impact on the distinguishability is very small for photons with long coherence times. However, the effect on the resonance condition is significant and generally needs to be compensated. This section will discuss two existing methods and introduce a new technique of compensation, and their effects on the cavity parameters and characterisation measurements. It is important to mention at this point that the compensation schemes are generally designed to compensate delays introduced over several round-trips, but as the exact position of pair creation inside the crystal is random, perfect compensation inside the cavity is not possible. Furthermore, we still need to adjust the temperature and possibly the alignment of the crystal in all cases to overlap the down-converted photons with the pump resonance in order to achieve triple resonance.

One way to compensate for the delay is shown in Fig. 3.3. By introducing a second birefringent crystal with its optical axis rotated by 90°, we can generate the same delay as the ppKTP, but for the orthogonal polarisation. The rotation switches the refractive indices $n_e \rightarrow n_o$ and vice versa and can be implemented by a KTP crystal without periodic poling to avoid phase matching and therefore stimulated emission. In an ideal case, the compensation crystal has exactly the same length as the ppKTP. This can be achieved in the manufacturing process by polishing both crystals end surfaces simultaneously. However, it will still be necessary to use temperature tuning to perfectly match the length of the cavity for both polarisation. Because temperature tuning is necessary in any case, we can use different lengths and overlap signal and idler resonances by setting the KTP temperature accordingly. In our experiments, we first utilised this technique with a 12.5 mm long KTP crystal for compensation. The additional crystal causes further losses and decreases the theoretical predictions for the finesse of the cavity to $\mathcal{F} = 266$, the FSR to $\nu_{FSR} = 243$ MHz due to the slightly longer cavity and therefore increases the linewidth to $\Delta\nu = 914$ kHz. Furthermore, the small cross section of 1 mm × 2 mm impedes the mode matching significantly, as the slightest misalignment results in higher losses due to clipping of the beam and hence increases the linewidth.

An alternative approach is the "flip-trick" we invented, using a half-wave plate (HWP) at 45° for compensation. The idea is as follows: a HWP inside the cavity changes the polarisation from horizontal (H) to vertical (V) and vice versa each round-trip. For now, we will only consider one polarisation mode, so one photon of the created pair, but all statements apply for the other photon as well. After being created, the photon is travelling along the ordinary axis of the crystal. When transversing the HWP, the photon moves orthogonally polarised through the rest of the cavity, hence, along the extraordinary axis of the crystal. After reaching the HWP

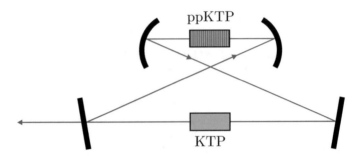

Fig. 3.3 Schematic setup for birefringence compensation via a KTP crystal. For simplicity, only one polarisation is shown. The KTP is rotated by 90° with respect to the other crystal, switching the ordinary and extraordinary axis and therefore allows to balance out the delay from the ppKTP. The compensation crystal should not be periodically poled to avoid stimulated emission

for the second time, the photon gets flipped back to the original polarisation mode. The accumulated time delay from the first crystal pass is compensated by travelling along the ordinary optical axis in the second pass. In general, all delays acquired in odd numbers of round-trips are cancelled out by passing the crystal in the orthogonal polarisation state in even round-trips. This process of delay and compensation is repeated until the photon leaves through the outcoupling mirror (or is lost). Note that the HWP ideally only affects the signal and idler wavelengths and leaves the pump light unchanged. A simplified scheme of the process is shown in Fig. 3.4, again, we picked only one starting polarisation for clarity.

This HWP "flip-trick" offers several advantages, but also some minor disadvantages compared to the compensation crystal method introduced before. We effectively double the length of the cavity for the single photons to two physical round-trips while leaving the pump light unaffected. Although the individual photons can still leave the cavity through the outcoupling mirror after any number of round trips, photon pairs are only detected as coincidences at even number of round-trip differences. After odd differences, one of the photons transverses the HWP an additional time, resulting in either HH or VV pairs. After the cavity, a polarising beam splitter deterministically separates orthogonal polarisations, consequently coincidences on individual detectors are only recorded for HV and VH pairs. In this compensation method, the outcoupling mirror simply takes the role of an extra source of loss every second round-trip.

Further proof that we actually double the length of the cavity and not only ignore the output at certain times is found in two different ways, with the corresponding results illustrated in Fig. 3.5. When scanning the cavity length and monitoring the resonance condition for the pump and single photon wavelengths on an oscilloscope, we expect the pump light to fulfil the resonance condition ($L = n\lambda$, $n \in \mathbb{N}$) twice as often as the down-converted light, if the cavity length is the same for both frequencies. Instead, we observe single photon resonances in both polarisations overlapping every pump resonance (Fig. 3.5a), demonstrating the different perimeters of the cavity. A

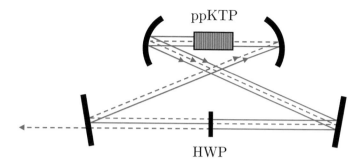

Fig. 3.4 Compensation method using a HWP at 45°, illustrated for one initial polarisation mode for clarity. After the photon is created inside the nonlinear crystal in a certain polarisation indicated by the solid line at the top left corner of the ppKTP, it travels through half the physical cavity length until it arrives at the HWP. Here, its polarisation gets flipped orthogonally (dashed line) and it transverses the crystal on a opposite optical axis, introducing a delay compared to the other polarisation mode (not shown). In a second round-trip the polarisation gets changed again and the formerly introduced delay is cancelled out, overlapping the resonances for signal and idler. Note: the spatial walk-off is only introduced for illustration purposes. In the experiment, orthogonal polarisations still travel along the same optical path

pleasant side effect of this is that the photon pair creation is enhanced independent of which pump resonance is used for stabilisation (see Sect. 3.4.2). The second way of confirming the new length of the cavity is measuring the distance between the resonances ($= \nu_{FSR}$) directly with respect to a frequency reference. We implement this measurement by scanning the frequency of the laser and using the light to probe the resonance condition of the cavity and Doppler-free rubidium D_1 transitions. The result is shown in Fig. 3.5b. We can find three different spectroscopy lines in rubidium within a scanning range of about half a GHz. The outermost peaks correspond to $S_{1/2}, F = 2 \rightarrow P_{1/2}, F' = 2$ and $S_{1/2}, F = 2 \rightarrow P_{1/2}, F' = 3$ transitions in [85]Rb, with a spacing of 362 MHz [27]. Calibrating the horizontal axis with this value results in a spacing of $\nu_{FSR} = 120.9$ MHz corresponding to a cavity length of 2458 mm. This is slightly more than the expected doubling of the perimeter, but well within the error margins of the single round-trip length measurement.

Another advantage of the HWP is that it does not increase the linewidth of the single photons, because the HWP itself is (almost) lossless. The additional losses, due to extra reflections, transversed surfaces and absorption, reduce the finesse to $\mathcal{F} = 161$, half of its original value without any compensation. However, as the FSR is halved concurrently, we preserve the original linewidth of $\Delta \nu = 755$ kHz.

A drawback of the method is that we lose around half of the brightness (coincidences) to the HH and VV photon pairs that are not separated at the polarising beam splitter and therefore not counted as coincidences. It also influences the characterisation measurements, e.g. $g_{s,s}^{(2)}(0)$, as the probability of accidental coincidences increases. A solution for this problem could be switching the polarisation of one of the photons after the cavity, conditioned on the arrival of the first one. In multimode operation, where the source emits many narrowband modes at different frequencies,

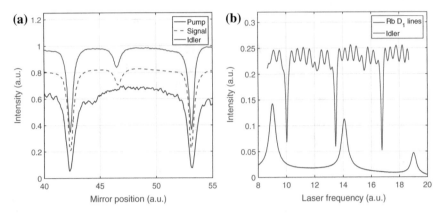

Fig. 3.5 Results of the measurements proving the effect of the HWP on the cavity length. (**a**) Measurement of the reflected cavity signal as a function of displacement of one mirror. We can clearly see that the resonance condition is fulfilled simultaneously for all light fields involved in the SPDC process. Adjacent resonances for the pump light are always overlapping with both polarisations of the single photons, proving the length variation at different wavelengths. Traces offset vertically for visibility. (**b**) Cavity and spectroscopy signal as a function of laser frequency. The peaks in the spectroscopy signal correspond to hyperfine transitions in ^{85}Rb, with the outermost lines separated by 362 MHz. With the precise knowledge of the spacing, the distance between neighbouring cavity resonances (FSR) in red can be calculate: $\nu_{FSR} = 120.9$ MHz

the arrival times of the photons are well-defined and a fast Pockels cell could flip the polarisation at times when HH or VV pairs are expected. Additionally, implementing this method before filtering allows to resolve the issue for the single-mode operation as well. A more technical problem is that theoretically the HWP should not interact with the pump frequency, but in practise it causes significant losses in certain circumstances, further discussed in Sect. 3.4.2. The effects may be reduced by adjusting the incoming polarisation and careful aligning the HWP itself, but cannot be balanced out completely.

The last method to compensate for birefringence effects is simple alignment of the ppKTP and its temperature. High control over all angular and translational degrees of freedom paired with high precision temperature control allows to tune small clusters, containing a few modes, on resonance, while other modes get suppressed. This method is combining birefringence compensation with an already frequency prefiltered output. Here, we will focus on the compensation aspects of the method, while the filtering is discussed in the Sect. 3.2.2.

In order to successfully implement this method in a triply resonant cavity, the available degrees of freedom have to be sufficient to overlap the resonances of all three fields involved, while offering enough birefringence to keep the number of modes per cluster small. The technique has been utilised for vastly different wavelengths of signal and idler [21, 28], small cavities of length similar to the crystal length [22, 29, 30] or a combination of both [31]. These implementations lead to a large tunability of the difference in FSR by small temperature adjustments within the phase-matching

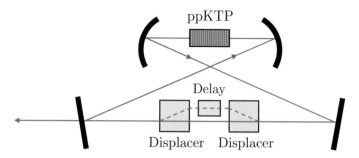

Fig. 3.6 Schematic cavity setup to implement the clustering effect. In order to keep the OPO lengthy but still introduce enough birefringence to achieve small clusters, we need to artificially increase the path length mismatch of the photon pairs. The two polarisation modes, indicated as solid and dashed red lines, overlap for most of the round-trip but eventually get separated by a beam displacer. One mode is then delayed and afterwards recombined with its partner on a second displacer. The remaining path is traversed collinear again

range. Realising the clustering effect in our setup is fairly challenging as we generate frequency degenerate photon pairs and our cavity is roughly 50 times larger than the crystal. A way to introduce enough birefringence and at the same time keep the physical cavity length at its current value is shown in Fig. 3.6. Signal and idler photons travel collinearly through the major part of the cavity, but are eventually separated in a beam displacer. One of the polarisation modes is then sent through a material of high refractive index to introduce the necessary delay. Afterwards, the photon pair gets recombined on another beam displacer and both photons propagate further in the same spatial mode.

Although the transmission of the displacers and the delay element is quite high (>99% each), the effect on the finesse and subsequently the linewidth is substantial. In case of the clustering effect, the FSR, finesse and linewidth are different for orthogonal polarisations: we can calculate a lower bound for the finesse to be $\mathcal{F}_S = 127$ and $\mathcal{F}_I = 159$, with equal losses for both polarisations in the displacers. Assuming a cavity length of 1219 mm again, the calculated FSRs are $\nu_{FSR,S} = 233$ MHz and $\nu_{FSR,I} = 236$ MHz. Finally, we can estimate the theoretical linewidths for signal and idler to increase to $\Delta\nu_S = 1.8$ MHz and $\Delta\nu_I = 1.5$ MHz, more than double compared to the HWP compensation method. The combination of the FSR with the approximate linewidth was chosen to result in only one dominant frequency mode per cluster and five clusters in the phase matching envelope. Further details on this method and the resulting mode structure can be found in Sect. 3.2.2.

Table 3.3 gives a summary of the impact of the compensation method on the FSR, the finesse and the linewidth of the photon pairs. We can see that the HWP method results in the lowest linewidth, however, on the expense of the brightness. The clustering effect would result in a very high coincidence rate per frequency mode, but the achieved linewidth will dramatically decrease the storage efficiency when combined with a quantum memory scheme based on sub-natural linewidths

Table 3.3 Boundaries of OPO parameters for different compensation methods

Compensation method	FSR (MHz)	Finesse >	Linewidth (kHz) <
Crystal	243	266	920
HWP	121	161	760
Clustering	233 (236)	127 (159)	1840 (1490)

like GEM [25, 26]. All results presented in this thesis will therefore benefit from the HWP as the compensation method.

3.2   Photon Pair Filtering

In Sect. 2.3.2 we have calculated the FWHM of the phase matching envelope for the ppKTP used in the experiments to be 100 GHz. This means that, together with a FSR of $\sim$121 MHz, the photons leaving the cavity have a spectrum consisting of approximately 800 narrowband modes, where the central degenerate mode has a creation probability of 0.12%. In order to efficiently interface the source with an atomic memory, further filtering is necessary to eliminate all undesired contributions to the spectrum and ideally create a single-mode, narrowband source. Due to the small spacing of 255 fm in the wavelength between adjacent modes, commercially available optical filters or Bragg gratings are not suitable for this application. Filters based on rubidium have decent mode extinction ratios and narrow transparency windows, but suffer from low throughput $\sim$10% [32, 33]. Optical cavities on the other hand can be designed specifically for this purpose to have very high transmission >80% and a likewise small transparency window. The window is reoccurring with a spacing equal to the FSR, but careful calculations paired with flexible cavity length adjustability can lead to a (nearly) single mode output. Another option is the clustering effect, briefly presented in Sect. 3.1.3 as a birefringence compensation method and here revisited as a spectral filter. The next sections present these two options to create degenerate photon pairs at the rubidium D_1 line and calculate the effects on the single photon spectrum.

3.2.1   Mode-Cleaning Cavity

For the mode-cleaning (filter) cavity we choose a monolithic, doubly resonant, triangular design. The cavity is free of birefringent elements and does not need a second focus for easier mode matching, making the triangular configuration a good choice in this case. The tilt of the in- and outcoupling mirror with respect to the light path in this architecture allows easy separation of stabilisation light and single photons.

The entire cavity is built out of one monolithic block of Invar and covered by an acrylic box for maximum temperature stability and protection against acoustic noise. The two flat mirrors are glued to angled sides at one end of the cavity and the curved mirror is mounted on a piezo-electric transducer at the other end for length stabilisation as shown in Fig. 3.7. We choose to stabilise the cavity length to the pump light at exactly double the frequency of the atomic transition. This allows easy filtering and spatial separation of the different wavelengths and enables continuous operation of the cavity, however, on the expense that only every second resonance of the pump light will allow transmission of the single photons.

The two light fields are coupled into the cavity through the flat angled mirrors from opposite sides. The counter-propagating travel directions inside the cavity further simplify the distinction between the single photons and the stabilisation light. The cavity length is controlled via the PDH technique by a Labview-driven FPGA (see Sect. 3.3.1), compensating length changes of the cavity with respect to a laser reference by adjusting the position of a piezo-mounted mirror. Due to the large angle of incident close to $45°$ for the light approaching the in- and outcoupling mirrors, double resonance is not automatically achieved and might change due to temperature fluctuations of the room on a daily basis. We overcome this issue by active temper-

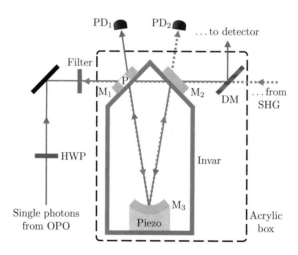

Fig. 3.7 Schematic design of the doubly resonant mode-cleaning (MC) cavity out of Invar with simplified optical paths. HWP, half-wave plate; Filter, narrowband filter for 795 nm; $M_{1,2}$, plane in-/outcoupling mirrors; P, Peltier elements; M_3, curved mirror mounted on piezo-electric transducer; $PD_{1,2}$, photo detectors; DM, dichroic mirror. The single photons are coupled into the cavity, while the cavity length is kept resonant with the blue (stabilisation) light at double the frequency. The error signal is generated by the stabilisation light, coupled backwards into the cavity, from PD_2 via the PDH method. The extra degree of freedom to control double resonance of the cavity is introduced by temperature tuning the incoupling mirror through three Peltier elements. The counter-propagating directions of travel automatically distinguish the stabilisation and single photon paths before, inside and after the cavity. A filter prior to the cavity suppresses backwards coupling of the blue light into the optical path of the photons while the dichroic mirror after the cavity allows to spatially overlap the stabilisation light to the single photon path. The monolithic design together with the acrylic box (with holes cut out for the optical paths) offers high temperature stability

ature control of the incoupling mirror (M_1 in Fig. 3.7) with three miniature Peltier elements glued directly onto the edge of the mirror. The capability to heat and cool the mirror provides the necessary second degree of freedom (apart from the length adjustment) to achieve double resonance.

Another effect of the angle between the mirrors is that the high-reflective coating varies significantly for orthogonal polarisations. This results in different finesses and linewidths of the cavity and subsequently changes the filtering capabilities drastically. The calculated and measured characteristic parameters of the mode-cleaning cavity for orthogonal polarisations are summarised in Table 3.4. The radius of curvature of the mirror combined with the length of the cavity determines the beam waist, important for mode matching of the incoming single photons. The FSR is the same for horizontal (H) and vertical (V) polarisation modes within the measurement accuracy. It is determined using rubidium transitions to calibrate the oscilloscope and averaged over multiple FSRs, similar to Fig. 3.5b. The linewidth is characterised in the same way and describes the FWHM of the Lorentzian-shaped resonance. The finesse is calculated using $\mathcal{F} = {\nu_{FSR}}/{\Delta\nu}$ and standard error propagation. From the finesse, we can determine the round-trip gain and further the losses inside the cavity to be 9.3% and 1.1% per round-trip for H and V polarisation, respectively. This is matching the theoretical calculations from the specifications of the three mirrors (9.4 and 1.1%) very well, confirming that there are no unaccounted sources of loss inside the cavity.

As the MC cavity solely accomplishes filtering of the photon pairs and does not affect their spectral width, the figure of merit here is the transmission on resonance. The cavity is outperforming filters based on rubidium [32, 33] by almost an order of magnitude on this important parameter. Assuming the photon linewidth is narrower than the cavity resonance (fulfilled in our case), the theoretically expected values for the throughput from Eq. 2.8 are 95 and 60%, both around 15% higher than the measured value in Table 3.4. This discrepancy arrises from additional losses on extra surfaces outside the cavity and, most importantly, the mode mismatch between the incoming light and the cavity mode. This mismatch leads to coupling into higher order spatial modes of the cavity that are not supported when kept on resonance with

Table 3.4 Mode-cleaning cavity parameters with uncertainties

Parameter (unit)	H polarisation	V polarisation
Radius of curvature of M_3 (mm)	250	
Beam waist ω_0 (μm)	161	
FSR ν_{FSR} (MHz)	766 (2)	
Round-trip length (mm)	391.5 (1.2)	
Linewidth $\Delta\nu$ (MHz)	11.9 (0.4)	1.35 (0.11)
Finesse $\mathcal{F}$	64 (2)	566 (45)
Round-trip loss (%)	9.3 (0.3)	1.10 (0.08)
Transmission on resonance (%)	80 (1)	46 (1)

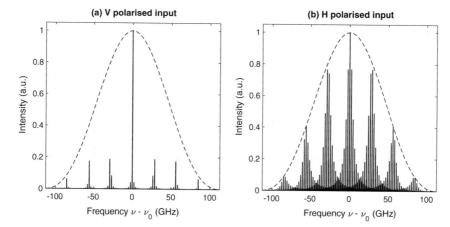

Fig. 3.8 Expected single photon spectrum after filtering by the MC cavity. Original phase matching envelope displayed as a dashed line. (**a**) Spectral filtering using the narrowband resonance for vertically polarised photons. The spectrum shows the dominant mode at ν_0 with all other modes in the phase matching envelope of the PDC significantly suppressed. (**b**) Broadband filtering utilising the horizontal polarisation mode. We still see elimination of the majority of the 800 modes, but the extinction performance is poorer than in (**a**), resulting in the necessity of an extra filtering step

the fundamental Gaussian mode, subsequently filtered out and causing the observed extra loss.

The calculated effect of the mode-cleaning cavity on the single photon spectrum for orthogonal incoming polarisation modes is shown in Fig. 3.8. There is a trade-off between the number of modes from the single photon spectrum after filtering and the throughput. If we chose the narrow V resonance, about 50% of the incoming signal is lost, but the mode spectrum shows only one dominant mode and a few small contributions after the cavity (Fig. 3.8a). On the other hand, high transmission around 80% is achieved for the H polarisation, but on the expense of five significant clusters in the spectrum as illustrated in Fig. 3.8b. This requires an additional filtering step, further reducing the brightness and adding more complexity to the setup. As the incident polarisation mode can be modified via a 3-paddle fibre polarisation controller, a polarising beam splitter and a HWP, almost effortless transformation of any incoming polarisation to pass the cavity vertically polarised is possible. This section showed that a carefully designed MC cavity can achieve very high suppression of a factor (ratio of unfiltered to filtered mode spectrum) of 360 for non-degenerate frequency modes. The throughput of about half of the photons at the desired frequency, restricted by fundamental and experimental coupling limitations, is still significantly outperforming other filtering systems.

3.2.2 Clustering Effect

Filtering the OPO output spectrum via the clustering effect is a fairly new technique that exploits the birefringence of the nonlinear crystal inside the cavity. By introducing a significant difference in the FSR of the two orthogonal polarisations, only small clusters of frequency modes are simultaneously resonant. As modes where merely one polarisation is resonant are suppressed, the output spectrum consist of only a few frequencies with their brightness significantly enhanced. As mentioned in Sect. 3.1.3, the method is usually implemented in short [22, 29, 30] or highly non-degenerate OPOs [21, 28], or a combination of the both [31]. In our experiment, the natural birefringence introduced by the nonlinear crystal leads to a very low mismatch $\Delta\nu_{FSR} = 122$ kHz on a FSR of $\nu_{FSR} = 244$ MHz at the degenerate mode, resonant with the rubidium D_1 line (795 nm). Additionally, temperature and angular tuning are generally not sufficient to accomplish triple resonance: the crystal dimensions of 1 mm $\times$ 2 mm $\times$ 25 mm prohibit big changes in the angles of the incoming light and the crystal length only accounts for a small fraction of the total cavity length, allowing very little temperature tuning of the resonances within the phase matching envelope. Assuming achievability of the resonance condition for the moment, the spectrum still consists of one cluster with >10 modes of considerable contribution as illustrated in Fig. 3.9a. Although the suppression factor is around 45, using only the natural birefringence is infeasible for our OPO and still requires an extra filtering step, disadvantageous for the brightness.

Alternatively, we can implement an OPO as shown in Fig. 3.6. Two barium borate ($\alpha - BBO$) beam displacers separate the orthogonal polarisations by $\sim$2.7 mm, enough space to introduce an extra BBO crystal in one of the arms without affecting the other. The extra temporal delay added to the path increases the FSR mismatch to $\Delta\nu_{FSR} = 3.7$ MHz. The BBO crystals are chosen because of their high refractive indices and reasonable losses below 1% per element. The extra loss increases the linewidths of the single photons to $\Delta\nu_S = 1.8$ MHz and $\Delta\nu_S = 1.5$ MHz, however, this is still well below the FSR mismatch as desired to obtain single-mode clusters. The spectrum of an OPO in this configuration consists of three main clusters consisting of one dominant mode each, shown in Fig. 3.9b. The frequency output is generated directly by the cavity without further filtering and thus achieves ultrahigh brightness. The suppression factor is 140, around 60% below the value of the narrow MC cavity filter.

The main advantages of the clustering effect, high photon creation rates and minimal spectral filtering without compromising the single photon linewidth, cannot be achieved in our current cavity design. Preliminary tests have shown the feasibility of the challenging cavity alignment procedure with the extra optical elements, especially the reliable recombination of the photon pair. The added optical components provide new controllable degrees of freedom (temperature, angle) to achieve triple resonance, making this technique a promising project for future implementation in the experiment.

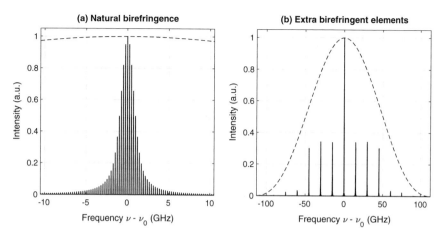

Fig. 3.9 Expected single photon spectrum for different implementations of the clustering effect. Original phase matching envelope displayed as a dashed line. (**a**) Zoom into the spectrum produced by the clustering effect from the SPDC crystal. The birefringence from the nonlinear crystal itself leads to only one cluster, but it includes several modes due to the small mismatch in FSR compared to the linewidth of the photons, requiring an extra filtering step. (**b**) Additional birefringence elements increase the FSR mismatch to 3.7 MHz, resulting in seven major, (almost) single mode clusters

3.3 Electronics and Control

So far, we have mainly discussed the optical elements and aspects of the setup to build an OPO and filter its output spectrum. An important part of the actual experiment generally tends to be forgotten: electronic components. They are the essential building blocks of all the control we have over the optical system, allow for probing and characterisation of optics and enable real-time evaluation of the recorded data. Considerate choices for each component need to be made to ensure smooth operation of the experimental apparatus.

This and the next section will discuss the soft- and hardware necessary to build the control loops, focussing on three different systems for control in place in our experiment: very powerful field-programmable gate arrays for cavity stabilisation circuits (Sect. 3.3.1), temperature controllers for crystals (Sect. 3.3.3) and frequency stabilisation for laser and cavities (Sect. 3.4). We will further introduce the software, an easy-modifiable code written in Labview, in place to control large parts of the experiment in Sect. 3.3.2.

3.3.1 Field-Programmable Gate Array

A field-programmable gate array (FPGA) is an integrated circuit which can be arbitrarily reconfigured by the customer for many different applications "in the field", hence the name. There are many books for people new to the field, e.g. [34, 35], and for advanced users [36, 37]. Easy reconfigurability, low energy cost and increased complexity are the main reasons for the success of this technology over the last decade. A large variety of applications, including signal processing, medical imaging, cryptography, computer hardware emulation and many more, are based on FPGAs. Their core consists of a number of logic blocks aligned in an array. The blocks can be rewired in different arrangements, usually by a combination of routing channels and I/O pads, to perform simple or complex calculations, so-called gates. The array usually includes some type of memory elements, like flip-flops and RAM, and oscillators for high precision clock generation, e.g. quartz-crystals. Especially the clock signal is crucial to synchronise the circuits. Modern FPGAs have further capabilities, for example common functions, hard-coded into the chip for increased processing speed and reduced required area.

In our experiment, we use FPGAs combined with frequency controllers, external clock generators and digital-to-analogue converters (DAC), all in one chassis by National Instruments (NI). This allows the creation of custom systems with onboard processing with the Labview software (also by NI). The FPGA fulfils the task of the necessary link between the incoming (outgoing) signals from (to) the setup and the control software. The control system is designed according to Ref. [38], allowing for eight frequency stabilisation loops simultaneously. This provides us with a high amount of flexibility to further expand the experiment if necessary. Using all components by the same manufacturer supports easy and efficient mapping of the code onto the FPGA circuit.

A simplified schematic of the physical system and the required hardware to implement PDH locking can be found in Fig. 3.10. The user interacts with the FPGA via a real time controller. The signal from a frequency generator regulated by the FPGA is divided in two parts, where one is sent to a clock to generate the modulation frequency for the sidebands and the other one is used to demodulate the AC signal derived from the reference system. The reference itself produces two signals: a slowly varying DC signal of a transition in an atomic vapour or a cavity resonance, and a fast AC signal containing the information from the sideband modulation that can be used to derive the error signal. Both are digitised and fed back to the FPGA, which then creates a control signal using a PII^2 control algorithm to stabilise the desired system. PII^2 have certain advantages over PID controllers, discussed e.g. in [39], however, as we do not use the I^2 part in our experiments (see Sect. 3.3.2), the theoretical treatment in Sect. 2.2.3 is sufficient.

Figure 3.10 illustrates the importance of the FPGA as the heart of the implementation of the stabilisation loops, interconnecting all its aspects: it controls the sideband frequency generation, demodulates the error signal, monitors the resonance or atomic transition, and creates the control signal to eliminate deviations from the reference. Additionally, we can faithfully create noise signals at every frequency below half the clock frequency at 80 MHz (defined in an adaptable look-up table) to disturb the lock and analyse its performance in different regimes. With this hardware part in mind, we can now look at the software that controls the FPGA.

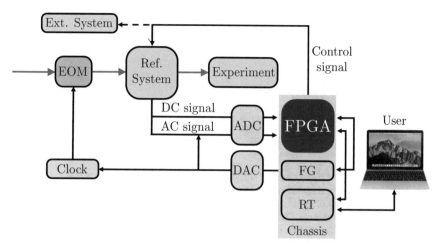

Fig. 3.10 Schematic design of the system controlled by the FPGA, adapted from [38]. Electric connections in black, optical paths in red. User, User interface; FPGA, field-programmable gate array card; FG, frequency generator; RT, real time controller; ADC, analogue-to-digital converter; DAC, digital-to-analogue converter; Clock, clock signal generator; EOM, electro-optical modulator; Ref./Ext. system, reference and external system; Experiment, remaining experiment. The user controls the FPGA via the RT. The FG generates the sideband frequency, modulated onto the light by an EOM and also used for demodulation of the fast AC signal coming from the reference system (cavity, spectroscopy). The DC signal determines whether the lock is successful and, dependent on the answer, switches between a scan and a lock mode. The FPGA uses the information from the AC and DC signal (digitised by the ADC) to create a control signal that is fed back to adjust the reference or an external system, dependent on the application

3.3.2 Control Software

The code to control the FPGA was written by our collaborators at the Australian National University and published in Ref. [38]. It was developed in the 32 bit version of Labview 2010 and therefore needs this version, or newer, to work. Labview allows easy modification and extension of the existing code to the needs of the user, without extended knowledge of programming languages. The control software is based on the PDH locking technique and uses an adjustable proportional, integral and squared integral (PII^2) algorithm for stabilisation. As previously mentioned, the code includes a noise generator to test the performance of the lock and a sequential locking logic for multiple consecutive stabilisation loops. This section will present how the software operates and briefly discuss known problems of the code.

The control software manages two different loops simultaneously: a fast loop is capable of handling high frequencies up to 40 MHz involved in the sideband modulation and error signal derivation, while a slow loop regulates signals below 750 kHz like cavity scans and control signals. Figure 3.11 illustrates the two regimes of operation and how they are connected. The code is capable of eliminating noise at frequencies within the slow loop regime, as this is where the control signal is

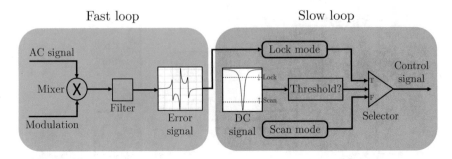

Fig. 3.11 Loops of the Labview control software. The AC signal is mixed down at the modulation frequency and then filtered to derive the PDH error signal in the fast loop. The slow loop compares the current value of the DC signal with a scan and lock threshold. A value below the lock threshold activates the lock mode that uses the error signal to stabilise the system to control. If the value rises above the scan threshold, the slow loop moves to scan mode using a sawtooth function until the DC value drops below the lock threshold again

generated. However, in our case this limitation is negligible because the main noise sources have low frequencies below 100 kHz. The system works as follows: the AC signal from the reference system is mixed down at the sideband modulation frequency in the fast loop. The resulting PDH error signal (see Sect. 2.2.1) is filtered and fed forward to the slow loop. Here, the reference DC signal is analysed and compared to the lock and scan threshold. Dependent on the DC value, the control system is either scanned past the resonance, using a sawtooth function, or stabilised to it, utilising the PII^2 control. When in the **AutoLock** setting, the control system will move from scan to lock mode when the DC value is below the lock threshold and vice versa when above the scan threshold.

Before running the code for the first time, a couple of checks have to be performed to make sure all elements operate according to their specifications. Most of these preliminary tests are related to correct names and addresses of all the devices and cards involved, with the desired information found in the Measurement and Automation Explorer (MAX) coming with Labview. Incorrect wiring or naming of any element in the code produces an error message and a log file, indicating the position and cause of the fault.

The control software has an integrated oscilloscope to monitor the AC and DC signal, the sawtooth scan and a forth port that can be configured by the user. All signals have separate multipliers and offsets to manipulate their representation as desired and are triggered to the sawtooth scanning signal. The code is capable of subsequently controlling eight different locks and all can be monitored individually on the oscilloscope. Problems with the oscilloscope are usually resolved by adjusting the acquisition rate or increasing the time out period, as discussed later in the section.

An example of the actual stabilisation control interface is given in Fig. 3.12. Throughout the whole program, bitshifts (**BS**) are coded into the software at various places, introducing multipliers (2^n) for fast value changes or to avoid saturation. The **Signal** column controls the fast loop, creating the modulation frequency,adjusting

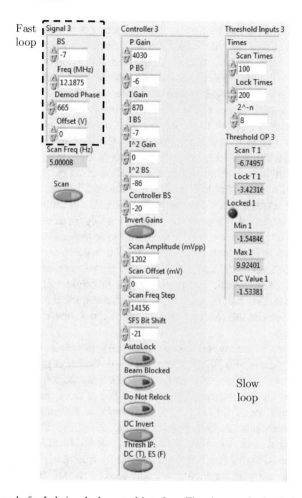

Fig. 3.12 Example for Labview lock control interface. The elements in the dashed rectangle in the top left corner control the fast loop: sideband modulation frequency, demodulation phase and possible offset for the error signal. It also contains the on/off switch for the sawtooth scan. The lower central column manages the parameters for this scan while the gains for the PPI^2 stabilisation algorithm can be changed in the top part. Finally, the right column sets the values for the scan and lock mode thresholds. Further details can be found in the text of Sect. 3.3.2

the demodulation phase and introducing an additional offset to the error signal if necessary. The modulation frequency has to be chosen according to the application. In our case, the lock to the rubidium spectroscopy uses sidebands around 1 MHz, limited by the bandwidth of the detectors in place. The cavity stabilisation loops have faster detectors, capable of recording the chosen 12.5 MHz modulation frequency. The strength of the modulation output signal can also be varied and is limited to ±1 V (not shown in the figure). In order to achieve optimal stabilisation performance, the demodulation phase needs to be adjusted to maximise the error signal. The exact phase of the maximum can be hard to find, but a practical way around this issue is to set the phase so that the error signal vanishes completely and add 90°, as minima and maxima are separated by $\pi/2$ due to the sinusoidal function used to create the sidebands. The demodulation phase may change between subsequent runs of the code, so this optimisation is performed each time the program is started.

The P, I and I^2 gain values can be adjusted in the top part of the **Controller** cluster, with an additional button to inverse the gain for convenience. The values need to be optimised for each lock separately using the Ziegler-Nichols (ZM) method introduced in Sect. 2.2.3. In general, the I^2 part of the algorithm is omitted as it introduces instabilities in most of our stabilisation circuits. At the beginning, the P and I gains are set to very low values to avoid oscillations that could cause the lock to become unstable instantaneously. We then slowly increase the **P gain** until oscillations become visible and set the value to roughly $^2/_3$ of that critical gain. Activating the **DC Invert** button should now result in the lock becoming more unstable, but if not, the switch needs to be set correctly. The same procedure is repeated for the **I gain** and alternating adjustments of P and I values are performed until a steady state is reached where the gains stop changing. In the original ZM method, the values are meant to be around half of the critical gain, but empirical investigations showed smaller residual deviations from the set point and higher robustness of the lock when using the factor $^2/_3$.

All parameters for the sawtooth scan (slow loop), amplitude, offset and frequency (function of step size times BS), can be changed in the lower part of the **Controller** column. The output for the scan is enabled via the **Scan** button and has a maximum range of ±10 V. Of the five switches at the bottom of the **Controller** column only two are of importance in our case: First, **AutoLock**, as mentioned before, turns the self-stabilisation loop on and off. Second, in the case of the rubidium spectroscopy lock, the DC signal as shown in Fig. 3.11 is upside-down. As the software is designed to minimise the DC signal (e.g. minimal reflectivity from a cavity), a signal inverter is coded into the program for such cases, activated by the **DC invert** switch.

Finally, the **Threshold Inputs** cluster on the right of Fig. 3.12 sets the proportionality factors for the threshold of the scan (**Scan Times**) and lock (**Lock Times**), with **Scan Times** $\leq$ **Lock Times**. The maximum, minimum and current values of the DC signal are shown in the **Threshold OP** tab. The actual values for the thresholds depend on the pre-factor 2^{-n} and the measured maximum and minimum DC values:

$$T_s = DC_{min} + \textbf{ScanTimes} \times 2^{-n} \times (DC_{max} - DC_{min}), \tag{3.2}$$

$$T_l = DC_{min} + \textbf{LockTimes} \times 2^{-n} \times (DC_{max} - DC_{min}). \tag{3.3}$$

Hence, the scaling values usually stay the same, independent of the DC signal strength but may vary between different locks. Higher values of the fractions result in a narrower range where the system is considered to be locked ("tighter" lock) and avoid stabilisation to any resonance aside from the main one. When activating the **AutoLock** switch, the system will perform one full scan to determine the minimum and maximum of the DC signal before trying to regulate. A common problem of the system refusing to go into lock mode is the demodulation phase being 180° out of phase. Switching on the **Invert Gains** button solves the problem without the need to change the phase or P and I gains. Further information on lock optimisation and trouble shooting can be found in Ref. [38].

Similar to any other type of software, the Labview-based control code suffers from a small number of known (and unknown) bugs listed below with their most likely solutions in brackets:

• Additional noise on error signals due to triggering failures from the frequency generator to the ADC (Restart Labview server, FPGA).
• Real time controller time-outs can cause the FPGA to be unresponsive to changes made in the user interface (Restart Labview server, FPGA).
• Clock generation issues can stop the FPGA from sending out the modulation signal (Reset clock or restart Labview server, FPGA).
• Scope triggering incorrectly/errors out (Change acquisition rate/increase time-out).
• Code not compiling due to lack of memory in FPGA (Delete unused code).
• Scan amplitude changes in 107 mV steps (Cannot be resolved, but step size was adjusted).
• Demodulation does not work properly (Change phase by 360° up or down).

3.3.3 Temperature Stabilisation

In order to achieve optimal phase matching for the SHG and SPDC process while keeping the cavity on resonance with the fields involved, a high degree of temperature control is necessary. The heart of the temperature stabilisation is a thermoelectric cooler (TEC), also known as Peltier element, regulated by a temperature controller. TECs use electrical energy to transfer heat from one (target) side of the device to the other, dependent on the direction of current flowing through them. Therefore, they can be used for heating and cooling of devices mounted, often glued, onto them. As the TEC only transfers but does not create heat, one side of the element becomes colder while the other side heats up. The increase or decrease of the target temperature needs to be compensated by a heat sink on the other side to keep the device from overheating/freezing.

Our nonlinear crystals are mounted in custom-designed holders out of copper, thermally glued onto a Peltier element for optimal heat transfer. The other side of the TEC is connected to a large aluminium heat sink placed on a 5-axis translation stage for control over all degrees of freedom of the crystal alignment. The stage itself is attached from underneath to the cavity through cut-outs, in order to protect the integrity of the Invar block. Figure 3.13 shows a 3D drawing of the whole mount, including the electric wires to control the TEC.

All temperature controllers in the setup are model LFI-3751 by Wavelength Electronics, specified to a temperature stability <2 mK. It is a high precision PID

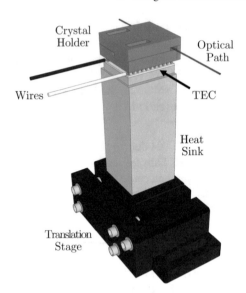

Fig. 3.13 3D model of the crystal mount. The TEC is thermally connected to the copper crystal holder on one side and the aluminium heat sink on the other. The translation stage guarantees full control over the crystal alignment when mounted in the cavity. The optical path of the laser through the crystal is perpendicular to the front surface of the crystal holder

controller with a maximal output current of ± 5 A and a maximum output power of 40 W. The device is preconfigured to operate with a TCS-610 thermistor and converts the measured resistance into temperature via the Steinhart-Hart equation [40]. An addressable RS-232 interface allows cascading of multiple controllers and remote access. The provided command and response packet structure simplifies controlling, monitoring and data taking via our own software, programmed in Linux.

We characterise the performance of the PID algorithm by first thermalising the mount with the environment and then record the actual temperature as a function of time when the controller output is activated. The response function is shown in Fig. 3.14a. The plot exhibits almost textbook-like behaviour: fast approach to the set temperature of 41.4 °C, only overshooting once and no oscillations. In order to evaluate the achieved temperature stability, the standard deviation σ_T of the measured temperature is calculated for 170 data points over six minutes. Dependent on the demand of accessibility, the OPO can be operated in different stability regimes with a variable amount of boxes enclosing the setup: when working on the alignment of the cavity, no cover boxes can be used as they would restrict the access to the mirrors. For the delicate adjustment of the crystal orientation we can install one acrylic box over the cavity and make all necessary changes to the translation stage from the outside. Finally, in the case of data taking, a second box encloses the breadboard carrying the whole setup to avoid short term thermal drifts caused by air currents from people walking through the laboratory or the air conditioning system. The influence of the boxes on the temperature deviation is illustrated in Fig. 3.14b.

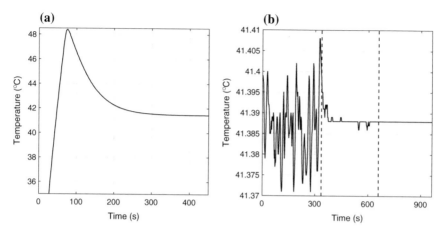

Fig. 3.14 Temperature controller characterisation. (**a**) Measured PID response starting at room temperature. The set temperature is reached and stabilised after around 10 min. (**b**) Effect of open setup (left), one box (centre) and double enclosure (right) on the temperature fluctuations. The first box around the cavity eliminates a vast majority of the fluctuations. After closing the second box around the entire setup the fluctuations are below the measurable accuracy level

Without additional protection from sudden air currents the temperature fluctuations are high, with $\sigma_T = 8.2$ mK. This might not sound significant, but the FWHM of the triple resonance of the OPO is on the same order ($\sim$10 mK). Therefore, the OPO cannot be operated emitting a constant rate of photons at triple resonance without extra thermal insulation form the environment. The doubly boxed system on the other hand exhibits very high temperature stability with $\sigma_T < 0.5$ mK, half of the smallest measurable temperature change. This is more than an order of magnitude below the uncovered performance and, in principle, allows constant operation on top of the narrow triple resonance.

Unfortunately, good temperature stability alone is not sufficient to ensure the triple resonance condition is fulfilled at all times. A second criterion is the difference between the actual (T_{act}) and the set temperature (T_{set}), combined with their fluent tunability. This represents the main problem in our experiment as it requires a high resolution analogue-to-digital converter inside the temperature controller. It turns out that, although the device can measure and set the temperature on a mK scale, it does not necessarily change the output according to every change in T_{set}. We performed excessive testing on multiple devices to characterise the "step size", the amount of change (dT_{set}) in the set temperature to cause a measurable change in T_{act} by the controllers. The step size hugely depends on the specific set temperature and the direction from which this temperature is approached. It ranges from the expected 10 mK, meaning a change in T_{act} for every change in T_{set} in this experiment, up to 140 mK, implying the controller only starts regulating after 14 changes in T_{set}. The effect is observed in all devices and exemplified in Fig. 3.15 for increasing and decreasing temperature changes. The plots show an abnormal operation behaviour, with a con-

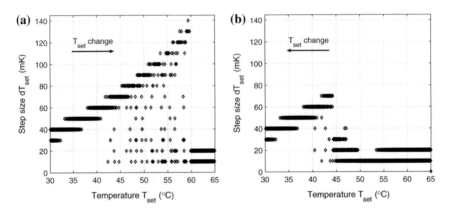

Fig. 3.15 Step size dependent on T_{set}. The threshold for the controller to react to a change in the actual temperature increases with the set temperature until a maximum is reached. At higher values the regulator behaves as expected, where (almost) every change in the input is followed by a corresponding change in the output. Threshold and step sizes depend significantly on whether T_{set} is (**a**) increased or (**b**) decreased and are roughly $dT_{set}(59.8\,°C) = 140$ mK and $dT_{set}(44.2\,°C) = 70$ mK, respectively

tinuous increase in step size to roughly 44 °C and 60 °C for negative and positive change of T_{set}, respectively. The exact reason for this phenomenon and its large hysteresis is unknown and could not be resolved with the manufacturer. A temporary solution to the issue is to slightly angle tune the crystal to triple resonance, causing phase-matching imperfections, collection path misalignment and subsequently decreasing the brightness. For future experiments, the controllers should be replaced with a custom-made stabilisation system to avoid this problem.

3.4 Frequency Stabilisation

Frequency stabilisation is crucial to ensure the compatibility of the single photon source with the atomic transition in frequency and linewidth. In our setup, there are multiple interconnected frequency stabilisation loops in place, all utilising the PDH technique and PID stabilisation algorithms introduced in Sect. 2.2. The laser frequency is locked to the length of the SHG cavity, which itself is controlled to be on resonance with the rubidium D_1 transition at 794.979 nm (377.107 THz) in vacuum [13]. The cascade structure ensures that the pump light created for the OPO has a narrow linewidth from the cavity as well as a frequency that is exactly half of the desired single photon frequency governed by the atomic lock. The SPDC cavity is stabilised to the incoming pump light, with the broad envelope of the down-conversion spectrum precisely centred around the degenerate cavity mode at the atomic line. The final step to achieve narrowband single-mode operation of the quantum source is the MC cavity, likewise resonant with the pump light as presented in Sect. 3.2.1. This last step is more for the purpose of frequency selection than actual stabilisation, but will still be included in this section. Here, we will describe the individual systems and the implementation of their stabilisation loops, while Sect. 3.6 gives and overview of their connections.

3.4.1 Laser

The main light source of the experiment is an amplified diode laser model TA Pro by Toptica. The emission has a typical grating-stabilised linewidth $\sim$100 kHz and is tunable from 780 to 800 nm, covering the D_1 and D_2 transition line in rubidium. The laser diode is set up in a Littrow configuration (see e.g. Refs. [41, 42]), where a grating is reflecting back a small amount of the emission spectrum into the diode, forcing it to emit at a single narrow frequency. The laser offers two fibre-coupled output channels: the master port is emitting around 15–20 mW of light directly from the diode, used to probe the absolute frequency reference. The slave port is enhanced by a tapered amplifier chip to a maximum of 2 W before being coupled into an optical fibre and sent to the experiment.

The system allows full control over current and temperature of the diode as well as the external grating alignment via a piezo-electric transducer to tune the output frequency as desired. The temperature and current of the amplifier chip has to be changed according to the input wavelength coming from the diode in order to achieve optimal amplification of the output mode. All control electronics come in individual modules combined in a single rack to simplify the replacement of faulty components. We have purchased the laser system with an additional Digilock 110 locking module, allowing the implementation of scans, PID controllers and frequency modulation together with a graphical user interface, described in the following paragraphs.

The software interface, shown in Fig. 3.16 with the transmission resonance and the pre-filtered error signal derived from the SHG cavity, offers a variety of options for monitoring, analysing and simulation of the locking performance. The sinusoidal modulation to derive the error signal has a tunable frequency between 17 Hz and 25 MHz in steps of approximately factor two, with a maximal amplitude of 2 V. The regulator bandwidth of the controller is $\sim$10 MHz, hence, we chose a modulation frequency of 12.5 MHz for a steep slope of the error signal and largest possible bandwidth. The signal is driving a resonant EOM to modulate the sidebands onto the carrier laser frequency for the PDH lock, avoiding the necessity to modulate the current of the laser diode. The resonant configuration has the advantage that we can use the signal generated by the Digilock without additional amplifiers that could introduce noise, but on the expense of reduced tunability. The user interface is equiped with an algorithm to automatically adjust the demodulation phase of the local oscillator to match the derivative of the spectrum signal, but can also be changed manually for fine tuning. The phase is adjusted correctly, if a large, symmetric error signal is visible on the display and a positive slope of the centre of the error signal

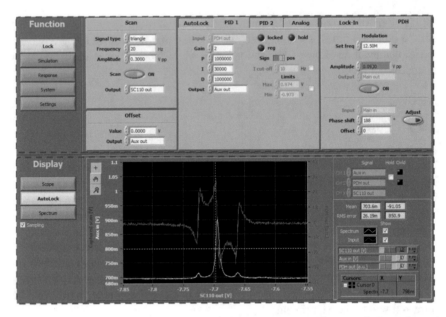

Fig. 3.16 Digilock user interface. The upper half of the screen is used for control, simulation and analysis of the PID locks. The lower part is displaying the cavity resonance in yellow, including the sidebands to both sides, and the corresponding PDH error signal in red. For further details see the description in the text

corresponds to an increase in frequency. Once the correct phase is found, the system can be stabilised to optimise the PID values.

The scan function allows to implement different signal types and adjustment for their frequency and amplitude. The signal is applied to the current controller and usually set to a triangular function at 20 Hz, with the magnitude (range) dependent on the application. Large scans are chosen for cavity alignment and coarse laser frequency tuning while laser locking is achieved with a small scanning range. The feedback loop consists of two PID controllers for frequency stabilisation: PID2 controls the Littrow grating via a piezo-electric transducer and co-dithers the current applied to the laser diode whereas PID1 only regulates the current. The integral gain of PID1 can be limited below a certain cut-off frequency to avoid the two controllers from accumulating offsets in opposite directions at low frequencies. Apart from that, both controllers are identical.

The lower part of the user interface is a two-channel oscilloscope and a spectrum analyser, displaying various signals derived from the inputs of the Digilock module. When in **AutoLock** configuration, the desired resonance feature and corresponding locking signal should be displayed within the oscilloscope as illustrated in Fig. 3.16, in order to choose the stabilisation set point with the crosshair (dashed line in oscilloscope). The Spectrum tab shows the fast Fourier transform (FFT) of the signal, transferring the **Scope** trace into the frequency domain. Refresh rate and averaging of the displayed signals can be modified as required in the settings section.

Finding the optimal settings for the gains of the PID controllers is not trivial, as both locks are feeding back onto the diode current and partially counteract each other if adjusted incorrectly. We implement an iterative trail-and-error method to find the optimal settings, starting with an unlocked laser with all PID values equal to zero for both controllers and increasing them slowly while monitoring the noise on the spectrum analyser and in the RMS error field. We first stepwise optimise PID2, starting with increasing the I gain until the system stabilises and then trying to minimise the low frequency noise spectrum below 25 kHz, the expected region of high amplitude noise. We alternate this optimisation on the I and P values until the noise amplitude cannot be minimised further, usually around half the value where an abrupt increase in the noise spectrum is observed due to oscillation of the stabilisation loop. We then apply the same method to PID1, always attempting to minimise the noise spectrum and regularly ensuring the laser is still locked. This is necessary, as a wrongly adjusted PID1 can easily disturb PID2 and cause the stabilisation to fail and furthermore, an unlocked laser can be misinterpreted as better stabilisation performance with less noise. Hence, regular confirmation of the locking status in the **AutoLock** oscilloscope setting is critical. After the best I and P values are found for both stabilisation loops, the D gain is adjusted to further minimise the noise if possible. Generally, the D gain allows higher I and P values by avoiding oscillations. In our experiment, it is expected and observed that the system is not very sensitive to these gains, especially the slower current feedback, as the importance of the differential response increases with frequency and high frequency noise has low amplitudes. We therefore set the D gain of PID1 to a well working value and decided to keep it at zero for PID2.

3.4.2 Cavities

Throughout our experiment, there are three cavities in place: one for SHG, one for SPDC and a third one for mode-cleaning. The design of the OPO and its features were already introduced in Sect. 3.1.2, and a detailed description of the triangular mode-cleaning cavity, its parameters and the expected impact on the spectra of the single photons is given in Sect. 3.2.1. Here, we will briefly introduce the SHG cavity and its measured parameters and discuss the role of both bow-tie resonators for frequency conversion and stabilisation in the experimental setup.

A 3D drawing of the SHG and SPDC cavity design including the mount for the nonlinear crystal introduced in Sect. 3.3.3, is shown in Fig. 3.17. The compensating element, only present in the SPDC cavity, is excluded for simplicity. In both cases, one of the curved mirrors is mounted on a piezo-electric transducer (PZT) for the implementation of the frequency stabilisation loops, allowing to scan or stabilise the cavity length. The main difference between the two resonators is that the SHG cavity is only resonant with the incident light at 795 nm, due to the almost transparent outcoupling mirror for the up-converted light ($R < 2\%$). Single resonance is sufficient in this case as the converted field is already high in intensity and SHG is inherently narrow in frequency if the pump laser is narrowband. Therefore, the cavity resonance is not needed to amplify and tailor the output light and we can avoid the necessity

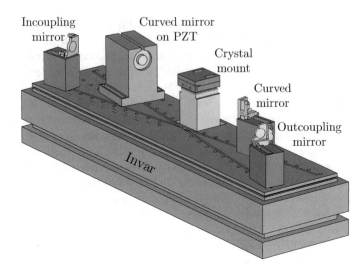

Fig. 3.17 3D drawing of the SHG and SPDC bow-tie design cavity for frequency control, conversion and narrowing, with the acrylic box cover not shown for simplicity. The distance between the four mirrors can be stabilised via the curved mirror mounted on a piezo-electric transducer (PZT). We are able to regulate the length of the cavity to the incoming light itself (SPDC) or an external atomic reference transition (SHG) via the Labview-controlled FPGA introduced in Sect. 3.3.2

Table 3.5 Measure parameters for SHG cavity and OPO with HWP compensation method, measurement uncertainties in curved brackets

Parameter [unit]	SHG, 795 nm	SPDC, 397.5 nm	SPDC, 795 nm
FSR [MHz]	278 (3)	241 (2)	121 (1)
Finesse	100 (2)	8.5 (0.3)	161 (4)
Linewidth [MHz]	3.0 (0.5)	28 (1)	0.75 (0.15)

of temperature tuning to double resonance on the expense of phase matching. The physical length of the SHG cavity is slightly shorter than the OPO. Additionally, the reflectivities for the incoupling/outcoupling mirrors vary significantly, leading to different values for the finesse and linewidths. Table 3.5 summarises the measured values for the FSR, finesse and linewidth of the SHG and the SPDC cavity. In case of the SPDC cavity, we determined the values for the single photon wavelength by probing with laser light at the target wavelength.

3.4.2.1   SHG Cavity

As briefly indicated in Sect. 3.4.1, the SHG cavity is used as the reference system for the laser, while simultaneously performing the frequency doubling to create the pump light for the OPO. This has multiple advantages: the resonant cavity enhances the conversion efficiency, as the laser light field inside the cavity is greatly amplified and transverses the crystal multiple times compared to the single-pass case. Furthermore, stabilising the laser to the resonator means that the generated SHG emission wavelength is also determined by the cavity length, due to the underlying nature of the frequency-doubling and linewidth-preserving SHG process. Subsequent implementation of a stabilisation loop to keep the cavity length itself resonant with the rubidium D_1 line (see Sect. 3.4.3), forces the created pump light frequency to be exactly double of the atomic transition. This is desirable, as the crystal inside the OPO is designed to have its peak emission at half the pump frequency, hence, back on resonance with the atoms. It is important to point out here, that the linewidth of the SHG given in Table 3.5 is also an upper bound for the blue light, but its actual value is much smaller ($\sim$100 kHz), governed by the laser linewidth. Building and stabilising the interconnected system is non-trivial and requires fast and robust locks, described below.

There are three major steps to achieve the necessary high level of noise suppression and stability of the feedback loop. First, we limit external influences to the system mechanically. The resonator is built on one block of Invar, with all mirrors mounted on large heavy posts (Fig. 3.17), and covered by an acrylic box to damp thermal air currents and acoustic disturbances, as previously discussed in Sects. 3.1.2 and 3.3.3. The second component is the correct adjustment of the PID control gains for the laser, presented in detail in Sect. 3.4.1. Here, we focus on the last step: the alignment of the nonlinear crystal has proven to be the final ingredient for optimal lock performance and conversion efficiency. Empirical investigations have shown, that larger SHG output power correlates with more effective frequency stabilisation, hence, we maximise the output to improve the lock.

In order to achieve a high conversion efficiency of the SHG, we adjust the orientation of the nonlinear crystal via the 5-axis translation stage, while the laser is stabilised to the cavity. This requires a robust lock to begin with, as adjustments of the crystal orientation introduce fast changes in the cavity length which the laser needs to be able to follow. The positioning procedure is performed by alternating changes of the horizontal and vertical position and angle of the crystal over a large amount of possible configurations, as several local maxima exist. We obtain an average coupling efficiency of 35% and SHG conversion efficiency of 45%, leading to an overall generation probability of $\sim$16% from the red laser light before the cavity into the blue frequency doubled field after. The stabilisation loop at this optimised crystal position shows maximal noise suppression and high robustness when analysed in the Digilock program. We believe it would be possible to further improve the total amount of created pump power by increasing the coupling, but generally we pump the OPO with <0.5 mW to remain far below threshold. While we are creating around 15 mW of blue pump light, the spare power is dumped, as an additional improvement of the locking performance has not been observed at higher powers. Therefore, more optimisation for extra conversion enhancement or control loop stability is not necessary in our experiment and contrarily increases the sensitivity of the output power to small day-to-day changes. Thus, most robust operation is achieved by the procedure above.

3.4.2.2 SPDC Cavity

The SPDC cavity is operated on triple resonance: signal, idler and pump field are resonant and enhanced simultaneously. This is far more demanding than the SHG case and requires precise control over temperature and alignment of the nonlinear crystal, as well as a high quality HWP for birefringence compensation. We stabilise the length of the cavity to the wavelength of the pump light at 397.5 nm via feedback to a PZT, overlap the signal and idler resonance with the HWP and finally match the different wavelengths by fine tuning the orientation and temperature of the crystal. The next paragraphs explain these steps in detail.

Initially, we implement the length stabilisation loop to the pump light, regulated by the Labview-controlled FPGA. Again, we use the PDH technique, modulating sidebands onto the pump frequency. We utilise an EOM at resonance frequency of 12.125 MHz, as the pump light still has residual sidebands at 12.5 MHz from the SHG process. Although suppressed in our cavity by $\sim$12 dB, these frequency components can interfere with the new modulation, causing instabilities in the demodulation phase of the error signal. The difference of 375 kHz between the sideband frequencies is the minimum step size of the FPGA. Additionally, the variation is well within the resonance bandwidth of 2.4 MHz of the EOM ensuring sufficient modulation depth and well outside the sideband linewidth of $\sim$90 kHz in order to cancel the interference effect.

The shape of the spatial mode after modulation strongly depends on the material of the EOM as various parameters like transparency or quality factor diminish, especially for frequencies close to the ultraviolet. Although we only operate at powers far below damage threshold on the order of mW, the initial EOM crystal, a magnesium oxide doped lithium tantalate (MgO:LT), leads to a high wavefront distortion of the spatial mode, as illustrated in Fig. 3.18. In order to observe the change, we take two snapshots of the mode profile on a camera after the EOM: without and with the EOM in operation. Starting at a beam diameter of $\sim$500 μm and an ellipticity (degree of deviation from a circle) of 0.94, we see that some type of cylindrical lensing effect alters the Gaussian beam profile, elongating the intensity along one dimension (ellipticity of $\sim$2.5) and causing multiple local maxima within around 30 seconds of operation. This makes efficient coupling into the OPO impossible and therefore the particular EOM crystal is unsuitable for our application. The issue was resolved by using a different material for the crystal, namely stoichiometric lithium niobate (sLN), that does not affect the shape of the spatial mode. The high transmission losses $\sim$40% are acceptable in our case, considering the small pump powers necessary to operate the triply resonant OPO.

The second step towards resonance is calibration and proper alignment of the HWP to overlap the degenerate signal and idler frequencies. The birefringence in the nonlinear crystal results in different cavity lengths for the orthogonally polarised single photons, described in detail in Sect. 3.1.3. A HWP inside the cavity compensates this effect while doubling the effective length of the cavity without causing extra losses for the single photon wavelengths. Ideally, the effect of the HWP on the pump polarisation or resonance should be negligible as a HWP at 795 nm is a

(a) **(b)**

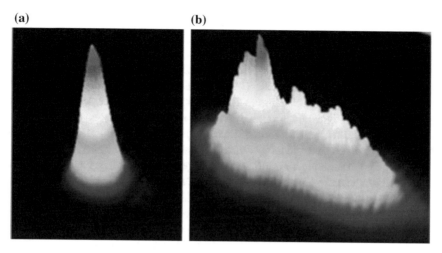

Fig. 3.18 Beam profile of the spatial mode of the blue pump beam after the EOM (**a**) without and (**b**) with the EOM operating. On the left, the mode emitted from the SHG cavity is (almost) perfectly Gaussian. The EOM seems to act as some type of cylindrical lens, stretching the spatial profile in one dimension and introducing multiple local maxima for the amplitude

"full-wave plate" at 397.5 nm and the thin optical element is expected to introduce only small extra losses. The first statement proves to be true: the polarisation of the incoming light is not altered by the HWP. However, the influence on the resonance via losses and consequently the coupling, finesse and linewidth are dramatic. The HWP offers four settings that lead to an orthogonally polarisation beam after transmission: $\Theta, \Theta', \Phi, \Phi'$. All of them affect the pump light differently. In the best alignment, the measured decrease in the finesse for the pump light from 17 to 8.5 indicates an increase in the loss per round-trip of 21%, caused by the HWP. According to Eq. 2.7, the higher loss roughly halves the achievable incoupling efficiency from 20.5 to 11%. Fortunately, the resonance is still sufficiently pronounced for locking purposes and the increased demand on incident pump light can easily be generated by the SHG system.

In order to fine tune the orientation of the HWP while satisfying the limited space restriction, we glued the HWP onto a half-inch high-precession manual rotation mount. This enables very fine alignment in a $\pm 7°$ range via an adjuster screw, where a 360° screw rotation corresponds to $\frac{4}{3}°$ change in the HWP position. In contrast to measurements of the cavity parameters presented earlier in this section, it is not possible to precisely probe the wave-plate alignment by shining red laser light backwards into the cavity. Due to the continuous-wave nature of the incoming light in this scenario, uncorrelated photons of both polarisations are constantly present in the OPO, leading to inconclusive results for the HWP optimisation. However, this method can still be used to align the collection paths when close to optimum for the HWP, as explained in Sect. 3.6. The necessary accuracy of the HWP angle within the cavity is achieved by looking at the difference in arrival times between the two orthogonal photons of a pair created in the OPO far below threshold, described in detail in Sect. 4.1.1. A classical characterisation of the HWP can be found in Appendix A.

Now that the cavity length is stabilised and both polarisations of the created single photons are compensated for birefringence, we need a last flexible degree of freedom to achieve triple resonance. Changing the crystal temperature is the perfect candidate as the geometrical light path through the cavity remains the same and we can control it externally, so direct interaction with the setup can be reduced to a minimum in order to lower air current and temperature fluctuations within the acrylic box around the setup. The exact temperature for the triple resonance changes on a daily basis, dependent on the room temperature. Still, the deviations within 24 h are never more than $50-70$ mK, well within the crystal phase matching temperature range of $1.7\,^{\circ}$C, and therefore temperature adjustments do not compromise the SPDC rate. Although the temperature stability achieved by the controller is <1 mK, well suited for the task, the poor resolution of the controller DAC does not allow the necessary accuracy, as discussed in Sect. 3.3.3. Due to the lack of a different degree of freedom, the crystal positioning has to be regularly optimised for triple resonance. This is more invasive than temperature changes and has the disadvantage that the beam path is altered and subsequently the collection system becomes misaligned, causing lower brightness. In general, this is the last resort and only used if the obtainable actual temperature is far away from triple resonance.

3.4.3 Atomic System

The atomic reference system in place uses Doppler-free spectroscopy [41] (see Sect. 2.2.2) in natural rubidium (Rb), a mixture of the two isotopes ^{85}Rb and ^{87}Rb, for absolute frequency stability of the SHG cavity and furthermore the laser and the OPO. We utilise a CoSy system by TEM Messtechnik, which contains the complete opto-mechanical spectroscopy setup and also the evaluation electronics to obtain the spectrum as an output signal on a BNC connector. A fibre-coupled laser beam can be connected to the pre-aligned structure, where it is separated into multiple paths passing through a rubidium vapour cell from different directions and is finally monitored on three photodiodes. The observed signals are internally amplified and processed to be displayed on an oscilloscope or for derivation of an error signal to stabilise the SHG cavity. The CoSy system is accompanied by an electric control module containing individual BNC connectors for different output signals and additional control elements for signal optimisation and cell temperature adjustment, shown in Fig. 3.19.

In order to achieve an optimal signal-to-noise ratio (SNR) of the spectroscopy and the error signal, the input beam needs to have a diameter of $2-3$ mm with $1-5$ mW of power and must be vertically polarised with reference to the table plane. In our fibre-coupled version of the system, the ideal diameter and alignment is set in the factory and cannot be changed. For a high SNR, an input power close to the minimum of 1 mW

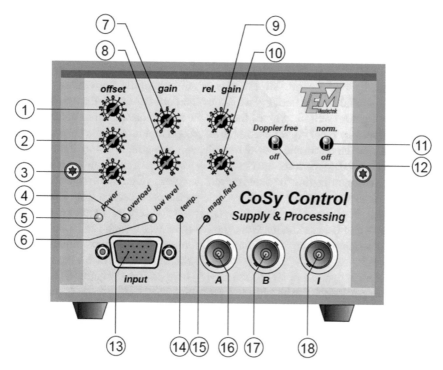

Fig. 3.19 Front panel of the CoSy electronic control unit, adapted from Ref. [43]. 1–3, dark current adjustments for channels A, B and I, respectively; 4–6, control lamps for electrical and optical power; 7–10, gain and relative gain adjustments for channel A and B; 11, switch to activate normalisation; 12, switch to activate Doppler-free spectroscopy; 13, input from optical CoSy system; 14 and 15, adjustment screws for temperature and magnetic field (optional); 16–18, BNC output ports. Further information on the purpose of each element can be found in the text

is selected to eliminate saturation effects causing excessive noise. The amplifier gain for the photodiodes can be chosen between 10^4 and $3.3 * 10^7$ V/A, with a factor of 3.3 between neighbouring positions. Its value is determined accordingly to the input power. The gain generally causes only small changes in the SNR of the spectroscopy signal, but the effect on the derived error signal can be significant. We therefore choose $3.3 * 10^4$ V/A, a value at the lower end of the range, in our experiment for a optimised SNR of the error signal. Finally, it is important to place the CoSy far enough from magnetised sources as they can interfere with the spectroscopy signal.

Although the optical system is pre-aligned and there is no option to change the path inside the CoSy enclosure, there are still a number of electronic adjustments that must be completed before stabilisation to a transition line is possible. The goal is to fine-tune the sensitivity and gains of the detection system to minimise electronic noise and derive a high contrast error signal. The next paragraphs give an instructional step-by-step overview to prepare the system for operation, with Fig. 3.20 illustrating measured examples of the signals after selected steps for the rubidium D_1 line [13, 27].

The CoSy offers three different output signals as shown in Fig. 3.19: A, B and I. Signal A and B display the spectroscopy while the I signal outputs a DC voltage, proportional to the measured laser intensity. The system is operational with a measured I value of $1-10$ V, but a voltage close to 1 V has proven to result in the highest SNR for the error signal. If the light level is appropriate, neither the **overload** nor the **low level** LEDs should be on. Initially, we adjust the dark current of the system. With no incident light present, the values for all three signals are meant to be zero. This can be achieved by blocking the incoming laser light and turning the corresponding **offset** knobs on the electronic control module, compensating for stray light and any possible photodiode dark currents. Next, the wavelength of the laser is set to match the atomic lines by adjusting the diode current and piezo voltage until transition characteristics on signal A and B become visible on the oscilloscope (Fig. 3.20a). The signal outside the spectroscopy features is adjusted to be zero by the **gain** parameters for channel A and B individually. Now, the **Doppler free** switch can be turned into the upper position, resulting in a weighted difference of the former similar signals of the two channels (Fig. 3.20b). The weighting can be changed with the **rel. gain** potentiometers until no Doppler broadening (A) or hyperfine structure (B) is visible. Lastly, the **norm.** switch can be activated, producing power-independent signals over a large range of input intensities (Fig. 3.20c).

Analogous to the other stabilisation loops in our experiment, the PDH technique is implemented on the system by sideband modulation through an EOM. Due to the lower bandwidth of the detectors inside the CoSy cell, the modulation frequency is limited to around 1 MHz. Therefore, the sidebands are within the width of the atomic resonance and not visible, e.g. in Fig. 3.20c. A higher frequency would allow a greater noise cancellation bandwidth, however, high frequency noise is small in amplitude and already canceled out by the laser stabilisation to the SHG cavity in our cascade implementation. For the control loop, the signal from channel A is split in two, with both halves filtered and amplified according to their application: in one arm, a low-pass filter generates a clear signal of the atomic stabilisation transition, while the other arm is bandpass-filtered to derive the error signal, only allowing frequencies around the modulation to pass.

It is important to mention that the spectroscopy and error signal are very sensitive to alterations of the incoming polarisation. Experience has shown that even the slightest changes in temperature of the EOM can have large effects on the polarisation, e.g. change a linear into an elliptical polarised beam. In order to reproducibly obtain a high SNR, the breadboard carrying the setup is therefore enclosed inside an additional acrylic box (see Sect. 3.6 for details). For full control over the polarisation we use a HWP in front and after the EOM in combination with a polarisation maintaining optical fibre to send the light to the atomic reference system. Nevertheless, using both wave plates to optimise the polarisation for spectroscopy is necessary on a daily basis.

The stabilisation loop itself is activated to control the length of the SHG cavity, while the laser is stabilised to the cavity. To avoid losing the laser lock while changing the length of the cavity, the necessary scan signal sent to the cavity PZT is applied at very low frequency of $\sim^1/_4$ Hz and small amplitude <200 mV, just sufficient to

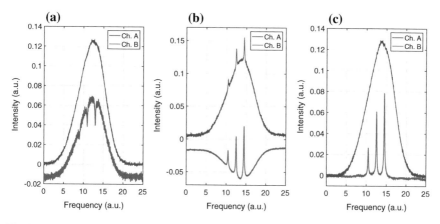

Fig. 3.20 Doppler-free spectroscopy signal alignment. (**a**) Initially, the narrow hyperfine spectrum of rubidium on both channels is thermally (Doppler-) broadened due to the velocity of the atoms in the vapour. (**b**) When activated, the signals from three internal photodiodes are electronically processed to derive a Doppler-free spectrum on Ch. A. (**c**) Further optimisation of the relative gains between the channels results in a background corrected saturation free spectroscopy signal on Ch. A (lower in red) and a signal free of the hyperfine structure on Ch. B (upper in blue), prepared to be analysed by the LabView software. For further details see corresponding text

cover the whole atomic transition. The method to determine the optimal gains for the PID loop is described in Sect. 3.3.2. The combined locking system is very robust against outside disturbances and remains continuously locked for 6−8 h.

3.5 Detectors

One of the fundamental tools for laser stabilisation and measurement of single photons are appropriate detectors. The requirements for these two applications are very different and can only be met by particular architectures. For PDH locking and continous-wave light monitoring, we use custom-made photodiode detectors that are fast and have a suitable bandwidth to resolve the modulation sidebands. In case of single photon counting, there are many options available: semiconductor-based detectors, transition edge sensors or superconducting nanowire detectors, just to name a few [44]. The results described in this thesis are all recorded by silicon single-photon avalanche detectors (SPAD), also known as Geiger-mode APDs, which offer high photon detection efficiency at the desired wavelength of 795 nm ($\sim$60%) and low dark count rates <25 Hz without the need of a vacuum, low temperatures or excessive expenditures. This section will first describe the home-made detectors for analysis of resonances and error signals and furthermore give a simplified introduction to APDs and their characteristic parameters in our experiment (PerkinElmer SPCM-AQR-14-FC) and the logic box for readout and processing (Roithner Lasertechnik TTM8000).

3.5.1 Universal Photodetectors

This section gives a brief overview of the most important electric elements and characteristic of our custom-made universal photodetectors (UPD), in place before and after every cavity to record the reflected light and help derive error signals, and monitor the transmitted power, respectively. Our devices are so-called transimpedance amplifier photodetectors [45], designed by collaborators at the Australian National University. They are based on operational amplifiers (op-amp) to provide gain to a voltage and/or convert currents into voltages, called transimpedance amplification, and photodiodes to convert optical power into an electric current. The components used in our UPDs have been chosen for their speed, excellent noise performance and high gain bandwidth product, determining the maximum gain and bandwidth of the feedback loop involving the op-amps.

There are two basic photodetector circuit that the op-amps are used in: a transimpedance amplifier first converts a current from a photodiode into a voltage before inverted amplifiers buffer the signal and provide a DC- and AC-coupled output, to be used to monitor steady powers or the sidebands, respectively. The two different circuits are illustrated in Fig. 3.21. The output voltage and gain of both amplifier designs is found by summing the input currents of the inverting port of the op-amp and is given as [46]

$$V_{out,i} = -\frac{R_f}{R_{in}} \times V_{in}, \tag{3.4}$$

$$V_{out,t} = -R_f \times I_p, \tag{3.5}$$

with R_{in} and R_f the incoming and feedback resistance, respectively, and I_p the photocurrent from the diode. The higher the gain of the circuit, the lower the maximal achievable bandwidth. Therefore, the amplifiers, especially the feedback resistor, need to be modified according to the application to ensure a stable feedback loop. A full circuit diagram of the detector can be found in Appendix B and further details on how to build the detectors are given in Ref. [46].

The photodiode in our detectors are characterised by their dark current and quantum efficiency. The diode converts an optical into an electric signal by, ideally, freeing an electron for every incoming photon. The real-life ratio of free electrons to incoming photons is defined as the quantum efficiency of the photodiode. The dark current describes the current present in the photodiode in the absence of light. It is mainly generated by background radiation and needs to be taken into account when we try to measure optical powers accurately. There is a large variety of suitable candidates commercially available, covering a wide range of wavelengths. We use silicon-based diodes model S5971 and S5972 by Hamamatsu Photonics [47] due to their high-speed response and highest quantum efficiency for the wavelengths in our experiment.

Probably the most crucial characteristic of the UPDs in our experiment is the noise spectrum. There are multiple possible causes for noise that need to be taken into consideration in order to achieve the precision needed. Typical noise sources

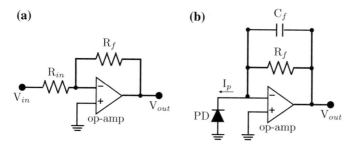

Fig. 3.21 Basic diagrams of the amplification circuits in our UPDs, adapted from [46]. (**a**) Inverted amplifier. The circuit inverts and amplifies an incoming voltage, corresponding to a multiplication with the factor $-R_f/R_{in}$. (**b**) Transimpedance amplifier. The design converts an input current (I_p) from the photodiode (PD) into an output voltage

are the photodiode, transimpedance stage and the op-amp. With the help of adequate software, it is possible to simulate the expected noise as a function of frequency. The contribution of the noise to the AC and DC output voltage for our sideband modulation frequency (12.5 MHz) is predicted to be <3 mV [46], far below our average signal strength on the order of 100s of mV.

3.5.2 Single Photon Counting

As of today, the standard technique of detecting single photons are silicon-based single photon detectors working in Geiger mode due to their small size, easy operation and low costs. The counting modules can generate a measurable avalanche current when triggered by one or several single photon(s) and are also able to record the arrival time to a sub-nanosecond accuracy. The basic simplified operation principle is as follows: a high electric field is applied to a p-n junction of a semiconductor. When a single charge carrier is injected in this high-field region, e.g. by photon impact ionisation, the particle gets strongly accelerated and generates secondary electron-hole pairs. This starts an avalanche-like multiplication process, growing exponentially and eventually limited at a constant macroscopic current level. The abrupt beginning of this current marks the arrival time of the photon. The device remains in this current-producing state and cannot detect further photons until it is quenched by an external circuit, lowering the electric field to stop the avalanche charge creation and allowing recombination of the electron-hole pairs. The time between rise and fall of the current is the characteristic dead time of the detector [48].

Processing of the generated signal from the single photon counting module (SPCM) requires a suitable electric circuit to quench the avalanche current and restore the diode for operation. Additionally, the circuit needs to detect the fast rising and falling edges of the current and convert it into a standard electrical output, in our case a rectangular transistor-transistor logic (TTL) pulse above 2.5 V for signal discrimination. The pulse is subsequently recorded by a time tagging module, storing the arrival time of each individual photon. We use software provided by the manufacturer to display and record the arrival times of single photons and calculate coincidence rates within variable time windows in real time. Final analysis of the raw data is performed by an easy modifiable code written in Matlab for full flexibility evaluation of the recorded events.

There are a number of basic performance parameters associated with single photon detectors, most prominently the detection efficiency, dark count rate and timing jitter. The photon detection efficiency is the probability that a single photon triggers a measurable avalanche current. It depends on optical properties such as wavelength and coupling as well as on electronic abilities related to the single photon rate or probability of creating a primary electron to start the avalanche. The detection efficiency strongly depends on the voltage applied to the semiconductor, increasing linearly at low voltages until saturation is reached [49]. The dark count rate describes the number of avalanches per second that occur mainly due to thermally induced charge carriers without an actual photon present. Its value is governed by the quality of the semiconducting material used in the diode, especially undesirable metal containment, and also increases with the applied voltage [49]. Hence, there is a trade-off between detection efficiency and dark counts dependent on the implemented electric field. The last important parameter is the timing jitter, a measure for the statistical fluctuations of the true arrival time of the photon. It arise from differences in the creation layers and transition times of charge carriers or noise in the electrical circuit [48].

We use a fibre-coupled, self-contained module by PerkinElmer (SPCM-AQR-14-FC) to detect the single photons at 795 nm. The device has a photon detection efficiency of 56% at the designated wavelength, well suited for many quantum optics experiments. The photodiode inside the module is temperature controlled to ensure stable performance independent of environmental fluctuations. The measured TTL pulse width of the detectors in our system is (35 ± 2) ns, with a timing jitter of (425 ± 75) ps. The device is specified to have a dead time of 50 ns, allowing single photon count rates up to 10 MHz. The decrease in detector efficiency at higher count rates due to more photons arriving within the detector dead time, is insignificant in our experiments as we usually operate at single photon rates below 100 kHz, where the decline is below 1%. The expected dark count rate of the modules is <100 Hz when all ambient light is blocked. In the actual experiment, this is not achievable as there will always be some stray light, e.g. from monitors, however, average measured dark count rates are <150 Hz, resulting in a very small contribution to coincidences between different channels, discussed in Sect. 4.1.

To analyse the electronic signal emitted from the SPCM, e.g. for correlations between different detectors, we need to further process the recorded single photon events. This is achieved by an additional counting logic in form of a time tagging module (TTM), in our experiment a model TTM8000 by Roithner Lasertechnik [50]. We can think of the device as a high resolution clock that logs the arrival time, input channel and direction of transition (low-high or high-low) of an external signal into an event table and sends it to a connected computer with suitable software installed.

Here, individual single photon count rates or histograms of arrival times between different channels are created in real time, allowing constant monitoring of the photon signals while e.g. changing the alignment of the collection path.

The TTM offers control of up to eight channels simultaneously and is fully configurable: recorded channel, direction of transition and the signal threshold can be individually adjusted to match the incoming TTL pulse from the detectors. The recorded time stamps have an accuracy of 82.3 ps ($=$ 1/clock frequency $=$ 1/12.15 GHz, hard-coded into the silicon circuit of the device), sufficiently below the detector jitter and therefore not limiting any measurements. Furthermore, individual time delays and filters can be implemented in the software on each channel, in order to compensate for external delays like different cable lengths or eliminate events without a neighbour in a certain time window for faster processing and smaller file sizes. All results presented in Chap. 4 are calculated through our own Matlab-based code in post-processing of the raw data stored by the included software.

3.6 Experimental Setup

This section gives a summary of the chapter, showing how and where the important components fit into the experiment and briefly explaining the feedback loops connecting them. We separate the description of the optical setup in three major parts, analogous to the actual experiment on the table which is built on three interconnected individual breadboards fulfilling different purposes: (1) laser and atomic frequency reference, (2) conversion breadboard including SHG cavity and OPO and (3) mode-cleaning setup, as shown in Fig. 3.22. The entire optical system fits on a single compact 1.2 m × 2.4 m table and is therefore well suited for transport to be merged with the GEM at the Australian National University (ANU) in Canberra. The experiment has moved laboratories twice in the last three years, resulting in a usual downtime of 2–3 weeks. The time delay to re-establish the source is mainly caused by problems with the temperature stabilisation and time necessary to plug in cables at the right connectors, not misalignment of optics. Although only small distances needed to be travelled each time, this already demonstrates the high portability of the system, required for its final journey.

3.6.1 Laser and Absolute Frequency Reference

The layout of the laser and CoSy breadboard is shown in the upper left corner of Fig. 3.22. Light from the master port of the diode laser is travelling through a polarisation-maintaining optical fibre into the sideband modulation setup. Before modulated, a small amount of the incoming light is separated on a beam splitter and fibre-coupled for further utilisation, e.g. as an alignment beam or for wavelength measurements. The remaining light transverses a HWP, a non-resonant EOM driven

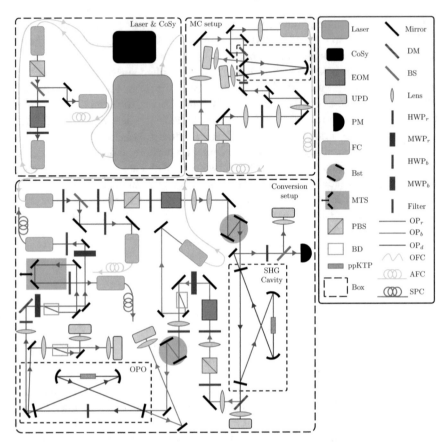

Fig. 3.22 Schematic representation of the complete optical setup of our experiment. MC setup, mode-cleaning setup; SHG cavity, second harmonic generation cavity; OPO, optical parametric oscillator. Legend: EOM, electro-optical modulator; UPD, universal photodetector; PM, power meter; FC, fibre connector; Bst, beam steerer; MTS, motorised translation stage, PBS, polarising beam splitter; BD, beam displacer; ppKTP, nonlinear crystal; Box, acrylic enclosure; DM, dichroic mirror; BS, beam splitter; HWP$_r$, half-wave plate for red light (795 nm); MWP$_r$, motorised half-wave plate for red light (795 nm); HWP$_b$, half-wave plate for blue light (397.5 nm); MWP$_b$, motorised half-wave plate for blue light (397.5 nm); Filter, 1 nm narrowband filter around 795 nm; OP$_r$, optical path red; OP$_b$, optical path blue; OP$_d$, dual optical path (red and blue overlapping); OFC, optical fibre connection; AFC, additional fibre connection (e.g. for alignment purposes); SPC, single photon connection (path to experiments). Arrows indicate the direction of light propagation. A detailed description on how the elements are connected is found in Sect. 3.6

at ~1 MHz and another HWP before it is sent into the CoSy atomic reference cell. As discussed in Sect. 3.4.3, the two HWPs are necessary to achieve full control over the polarisation for the EOM and the CoSy, both very sensitive to polarisation misalignment. The second (amplified) output port of the laser is also fibre-coupled and sent to the conversion setup.

3.6.2 Conversion Setup

The breadboard carrying the conversion cavities is the heart of the experiment, taking up roughly half of the optical table as depicted in Fig. 3.22. The light from the amplified output port of the laser is sent into the experiment, manipulated with HWPs, modulated by an EOM and finally coupled into the SHG cavity with a beam steerer. The beam steerer is a compact device consisting of two mirrors on a single mount that gives us control over the translational and angular degrees of freedom of the incoming light in order to simplify the overlap with the spatial cavity mode. Alignment of the position and diameter of the beam focus is achieved by a telescope, consisting of three lenses in front of the cavity. To monitor the laser and subsequently have an estimate of the SHG power, a portion of the back reflected signal from the frequency-doubling cavity is detected on a power meter. This allows us to constantly observe the incident power and coupling efficiency when scanning and locking the laser, respectively. The SHG resonator is described in Sect. 3.4.2.

After the cavity, where both wavelengths are present, the light is separated on a dichroic mirror: the red light is detected on a UPD and analysed in the Digilock program while the blue light is sent towards the OPO. The combination of a HWP mounted in a motorised rotation stage with a polarisation sensitive beam displacer enables us to fine tune the amount of power we send to the OPO on a μW scale. The unused blue light is fibre coupled and sent to the MC cavity for stabilisation purposes. Similar to the SHG cavity, we use a beam steerer plus a three lens telescope for spatial mode matching of the bow-tie SPDC cavity. Further information on the design and specifications of the OPO are discussed in detail in Sect. 3.1.

After the single photon cavity, the down-converted photon pair propagates collinear until split on a polarising beam displacer and collected via fibre-couplers. One of the photons can be delayed variably by a pair of mirrors, mounted on a motorised translation stage, to match the path length of its orthogonal partner. Efficient collection of the light particles into optical fibre is non-trivial. In order to align the collection path, continous-wave laser light from one of the additional ports is sent backwards through the single photon collection arms and the optical paths are overlapped on the beam displacer. The orthogonally-polarised signals, back-reflected from the OPO, are further detected on UPDs and displayed on an oscilloscope to optimise the mode-matching to the OPO. This does not necessarily guarantee ideal collection of the single photons but serves as a good tool for rough alignment. Final fine tuning is achieved by monitoring the single photon count rates on the SPCMs while slightly adjusting mirrors and collimation lens positions. From here, the fibre-coupled photons can either be detected straight away or first filtered in the mode-cleaning setup, if single-mode operation of the source is necessary as described below.

3.6.3   Mode-Cleaning Setup

For experiments requiring single-mode operation of the source, most prominently integration of the source in the GEM setup at the ANU, the mode-cleaning setup is essential. Shown in the upper centre of Fig. 3.22, it includes all the optics and electronics to stabilise the MC cavity to the SHG light at 397.5 nm and is optimised to achieve the highest possible transmission at the single photon wavelength, previously described in Sect. 3.2.1. Optimal transmission is achieved by very careful designing the light path between the incoupler and the fibre connector at the end of the table and making all alignment degrees of freedom adjustable at any position of the setup. The restrictions on coupling of the pump frequency are not as stringent, as the signal is solely utilised to stabilise the length of the cavity. A dichroic mirror behind the cavity separates the light at 795 and 397.5 nm in direction of travel of the single photons and overlaps them along the optical path of the pump light. The cavity is enclosed by an acrylic box for improved temperature stability, with small holes allowing unimpeded optical access.

3.6.4   Feedback Loops

Section 3.4 has described all frequency stabilisation components of the system in place and introduced how they fit in the setup. Here, we briefly summarise and illustrate (Fig. 3.23) how the regulator loops are connected to each other. The laser

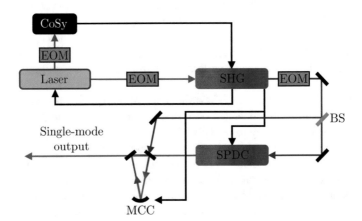

Fig. 3.23 Simplified drawing of the optical paths and electric feedback loops of the most prominent elements in the setup. Optical paths in red and blue, electronic feedback in black. CoSy, atomic frequency reference; SHG, second harmonic generation cavity; SPDC, spontaneous parametric down-conversion cavity (OPO); MCC, mode-cleaning cavity; EOM, electro-optical modulator for sideband creation; BS, beam splitter. Description in text

is stabilised to the SHG cavity by feedback onto the laser diode current and a piezo-electric transducer (PZT) to change the angle of a Littrow grating. The cavity itself is stabilised to the atomic transition by adjusting its length through a mirror mounted on a PZT. The interconnected loop ensures the creation of light at exactly double the frequency of the rubidium D_1 transition to further pump the OPO. A mirror mounted on a PZT controls the length of the SPDC cavity to be resonant with the incoming pump light. Finally, the single photons are filtered to have a single-mode output in the mode-cleaning cavity, also controlled to be at resonance with the pump light at 397.5 nm. All stabilisation loops use sideband modulation via EOMs, the PDH technique to create the error signals and PI(D) algorithms for feedback, described in Sect. 2.2.

References

1. Gisin, N., Ribordy, G., Tittel, W., Zbinden, H.: Quantum cryptography. Rev. Mod. Phys. **74**, 145–195 (2002)
2. Takemoto, K., et al.: Quantum key distribution over 120 km using ultrahigh purity single-photon source and superconducting single-photon detectors. Sci. Rep. **5**, 14383 (2015)
3. Agrell, E., et al.: Roadmap of optical communications. J. Opt. **18**, 063002 (2016)
4. Korzh, B., et al.: Provably secure and practical quantum key distribution over 307 km of optical fibre. Nat. Photonics **9**, 163–168 (2015)
5. Ten, S.: Ultra low-loss optical fiber technology. In: 2016 Optical Fiber Communications Conference and Exhibition (OFC), pp. 1–3 (2016)
6. Briegel, H.-J., Dür, W., Cirac, J.I., Zoller, P.: Quantum repeaters: the role of imperfect local operations in quantum communication. Phys. Rev. Lett. **81**, 5932–5935 (1998)
7. Chou, C.-W., et al.: Functional quantum nodes for entanglement distribution over scalable quantum networks. Science **316**, 1316–1320 (2007)
8. Duan, L.M., Lukin, M.D., Cirac, J.I., Zoller, P.: Long-distance quantum communication with atomic ensembles and linear optics. Nature **414**, 413–418 (2001)
9. Sangouard, N., Simon, C., de Riedmatten, H., Gisin, N.: Quantum repeaters based on atomic ensembles and linear optics. Rev. Mod. Phys. **83**, 33–80 (2011)
10. Tittel, W., et al.: Photon-echo quantum memory in solid state systems. Laser Photonics Rev. **4**, 244–267 (2010)
11. Yuan, Z.-S., et al.: Experimental demonstration of a bdcz quantum repeater node. Nature **454**, 1098–1101 (2008)
12. Munro, W.J., Stephens, A.M., Devitt, S.J., Harrison, K.A., Nemoto, K.: Quantum communication without the necessity of quantum memories. Nat. Photonics **6**, 777–781 (2012)
13. Steck, D.A.: Rubidium 87 D line data (2015). http://steck.us/alkalidata/
14. Chua, S.S.Y., et al.: Impact of backscattered light in a squeezing-enhanced interferometric gravitational-wave detector. Class. Quantum Gravity **31**, 035017 (2014)
15. Boyd, G.D., Kleinman, D.A.: Parametric interaction of focused gaussian light beams. J. Appl. Phys. **39**, 3597–3639 (1968)
16. Targat, R.L., Zondy, J.-J., Lemonde, P.: 75%-efficiency blue generation from an intracavity ppKTP frequency doubler. Opt. Commun. **247**, 471–481 (2005)
17. Scholz, M., Koch, L., Benson, O.: Statistics of narrow-band single photons for quantum memories generated by ultrabright cavity-enhanced parametric down-conversion. Phys. Rev. Lett. **102**, 063603 (2009)
18. Wolfgramm, F., de Icaza Astiz, Y.A., Beduini, F.A., Cerè, A., Mitchell, M.W.: Atom-resonant heralded single photons by interaction-free measurement. Phys. Rev. Lett. **106**, 053602 (2011)

19. Zhou, Z.-Y., Ding, D.-S., Li, Y., Wang, F.-Y., Shi, B.-S.: Cavity-enhanced bright photon pairs at telecom wavelengths with a triple-resonance configuration. J. Opt. Soc. Am. B **31**, 128–134 (2014)
20. Pomarico, E., Sanguinetti, B., Osorio, C.I., Herrmann, H., Thew, R.T.: Engineering integrated pure narrow-band photon sources. New J. Phys. **14**, 033008 (2012)
21. Fekete, J., Rieländer, D., Cristiani, M., de Riedmatten, H.: Ultranarrow-band photon-pair source compatible with solid state quantum memories and telecommunication networks. Phys. Rev. Lett. **110**, 220502 (2013)
22. Monteiro, F., Martin, A., Sanguinetti, B., Zbinden, H., Thew, R.T.: Narrowband photon pair source for quantum networks. Opt. Express **22**, 4371–4378 (2014)
23. Rambach, M., Nikolova, A., Weinhold, T.J., White, A.G.: Sub-megahertz linewidth single photon source. APL Photonics **1** (2016)
24. Steinlechner, J., et al.: Absorption measurements of periodically poled potassium titanyl phosphate (PPKTP) at 775 nm and 1550 nm. Sensors **13**, 565 (2013)
25. Hosseini, M., Sparkes, B.M., Campbell, G., Lam, P.K., Buchler, B.C.: High efficiency coherent optical memory with warm rubidium vapour. Nat. Commun. **2**, 174 (2011)
26. Hosseini, M., Campbell, G., Sparkes, B.M., Lam, P.K., Buchler, B.C.: Unconditional room-temperature quantum memory. Nat. Phys. **7**, 794–798 (2011)
27. Bowie, J., Boyce, J., Chiao, R.: Saturated-absorption spectroscopy of weak-field zeeman splittings in rubidium. J. Opt. Soc. Am. B **12**, 1839–1842 (1995)
28. Rieländer, D., Lenhard, A., Mazzera, M., de Riedmatten, H.: Cavity enhanced telecom heralded single photons for spin-wave solid state quantum memories. New J. Phys. **18**, 123013 (2016)
29. Chuu, C.-S., Harris, S.E.: Ultrabright backward-wave biphoton source. Phys. Rev. A **83**, 061803 (2011)
30. Ahlrichs, A., Benson, O.: Bright source of indistinguishable photons based on cavity-enhanced parametric down-conversion utilizing the cluster effect. Appl. Phys. Lett. **108**, 021111 (2016)
31. Luo, K.-H., et al.: Direct generation of genuine single-longitudinal-mode narrowband photon pairs. New J. Phys. **17**, 073039 (2015)
32. Cerè, A., et al.: Narrowband tunable filter based on velocity-selective optical pumping in an atomic vapor. Opt. Lett. **34**, 1012–1014 (2009)
33. Predojević, A., Mitchell, M.: Engineering the atom-photon interaction: controlling fundamental processes with photons, atoms and solids. In: Nano-Optics and Nanophotonics, 1st edn. Springer International Publishing (2016)
34. Maxfield, C.: FPGAs: Instant Access, 1st edn. Elsevier Science (2011)
35. Meyer-Baese, U.: Digital Signal Processing with Field Programmable Gate Arrays, 4th edn. Springer Publishing Company, Inc. (2014)
36. Kilts, S.: Advanced FPGA Design: Architecture, Implementation, and Optimization, 1st edn. Wiley-IEEE Press (2007)
37. Stavinov, E.: 100 Power Tips For FPGA Designers, 1st edn. CreateSpace, Paramount (2011)
38. Sparkes, B.M., et al.: A scalable, self-analyzing digital locking system for use on quantum optics experiments. Rev. Sci. Instr. **82** (2011)
39. Cho, J., Chong, S.: Stabilized max-min flow control using pid and pii2 controllers. In: Global Telecommunications Conference, 2004. GLOBECOM'04, vol. 3, pp. 1411–1417. IEEE (2004)
40. LFI-3751 with Autotune PID Thermoelectric Temperature Controller. Technical Report, Wavlength Electronics, Inc. (2003)
41. Demtröder, W.: Laser Spectroscopy: Basic Concepts and Instrumentation. Advanced Texts in Physics, 5th edn. Springer, Berlin, Heidelberg (2007)
42. Paschotta, R.: Field Guide to Lasers. Field Guide Series, Society of Photo Optical, 1st edn (2008)
43. User manual for CoSy 4.0. Technical Report, TEM Messtechnik GmbH (2006)
44. Eisaman, M.D., Fan, J., Migdall, A., Polyakov, S.V.: Invited review article: single-photon sources and detectors. Rev. Sci. Instr. **82**, 071101 (2011)
45. Horowitz, P., Hill, W.: The Art of Electronics, 2nd edn. Cambridge University Press, New York (1989)

46. Stefszky, M., Gmeiner, J.: ANU Photodetector V4. Technical Report, Australian National University, Canberra, ACT, Australia (2014)
47. Si PIN photodiodes. Technical Report, Hamamutsu Photonics K.K., Solid State Devision (2015)
48. Cova, S., Ghioni, M., Itzler, M.A., Bienfang, J.C., Restelli, A.: Chapter 4-semiconductor-based detectors. In: Alan Migdall, J.F., Sergey, V.P., Bienfang, J.C. (eds.) Single-Photon Generation and Detection. Experimental Methods in the Physical Sciences, vol. 45, pp. 83–146. Academic Press (2013)
49. Ghioni, M., Gulinatti, A., Rech, I., Zappa, F., Cova, S.: Progress in silicon single-photon avalanche diodes. IEEE J. Sel. Top. Quantum Electron. **13**, 852–862 (2007)
50. TTM8000-time tagging module with 8-channels. Technical Report, Roithner Lasertechnik GmbH (2015)

Chapter 4
Single Photon Characterisation

Single photons are essential building blocks for quantum information and quantum communication, due to their high mobility, simple encoding and low interaction with the environment. In order to compare individual sources from the same or different architectures, we defined multiple metrics to characterise the single photons in Sect. 2.4. It is important for the community in this research field, to agree on a common ground for the measurement techniques to derive these numbers in order to make their comparison easy and meaningful. The particular features can be divided in two categories: classical and quantum characteristics. Spectral brightness, linewidth or FSR can all be described without quantum mechanics and are therefore considered classical. However, it is important to point out that they still require individual single photon and coincidence detection to be measured in the experiments. Quantum features, e.g. an auto-correlation value <1 or Hong-Ou-Mandel interference, on the other hand, can only be explained by quantum physics. Section 4.1 will present the classical results and the path to obtain them, while Sect. 4.2 deals with the quantum characteristics of the created photon pairs.

4.1 Classical Characterisation

4.1.1 Intensity Cross-Correlation Function $G_{s,i}^{(2)}(\tau)$

Measuring the intensity cross-correlation function is a convenient way to determine important characteristics of the OPO, for example the linewidth, with high accuracy. In general, $G_{s,i}^{(2)}(\tau)$ uses information from the single photon arrival times to determine the temporal shape of their wave function. The experimental setup to measure the auto-correlation function is fairly simple and shown in Fig. 4.1. The fibre-coupled single photons from the OPO are detected on two separate SPCMs, with the

© Springer Nature Switzerland AG 2018
M. Rambach, *Narrowband Single Photons for Light-Matter Interfaces*,
Springer Theses, https://doi.org/10.1007/978-3-319-97154-4_4

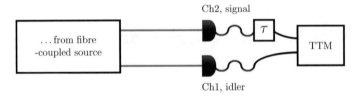

Fig. 4.1 Schematic of the experimental setup to measure the intensity cross-correlation function $G_{s,i}^{(2)}(\tau)$. The fibre-coupled photons from the SPDC cavity are detected on two separate SPCMs and their arrival times are analysed for coincidences between the two channels by a time tagging module (TTM). The signal arm is electronically delayed by τ to obtain the statistics for $G_{s,i}^{(2)}(\tau)$

signal arm electronically delayed by τ, and analysed to find coincidences between the two channels. The specific shape of $G_{s,i}^{(2)}(\tau)$ in our case of cavity-enhanced SPDC is defined by three parameters: the cavity decay rate, and FSR, for the individual photons, as well as the phase-matching envelope of the nonlinear crystal. The combination of the latter two gives an estimate of the number of modes contained in the OPO spectrum. As discussed in Sect. 3.2, the FWHM of the envelope is 100 GHz which, together with a FSR of ∼121 MHz, leads to about 800 modes. The upper limit of the cavity decay rate is calculated in Sect. 3.1.3 to be 760 kHz. Therefore, all parameters necessary to describe the cross-correlation function (see Sect. 2.4.2) are given from theoretical predictions and this section presents the related experimental results. However, the first step towards the final results is fine tuning the alignment of the HWP inside the OPO to perfectly switch between orthogonal polarisations every round-trip.

4.1.1.1 Half-Wave Plate Alignment

Detailed characterisation of the intra-cavity HWP is achieved similar to other classical parameters (e.g. the linewidth) by measuring the $G_{s,i}^{(2)}(\tau)$ and interpreting its results. Small rotational steps around $1/3°$ of the HWP between consecutive experiments, while constantly monitoring the results, enables us to determine the angle deviation from optimum with accuracy better than 0.2°. The experimental results are shown in Fig. 4.2. The plots illustrate the normalised number of coincidences (without background correction) between the two orthogonally polarised signal and idler photons as a function of delay in their arrival times (τ), or in other words: the cross-correlation function $G_{s,i}^{(2)}(\tau)$. The titles of the different sub-figures indicate the individual relative position of the HWP, with 0° being the angle after a preliminary coarse adjustment to 72° (see Appendix A). The data points are connected with lines for increased visibility of the comb-like structure, which arises from the higher probability of detecting signal and idler photons at multiples of effective cavity round-trip times t_{rt}. For an ideally adjusted angle ($-2°$ in Fig. 4.2), the HWP rotates horizontal (H) to vertical (V) polarisation and vice versa at every pass and the distance between

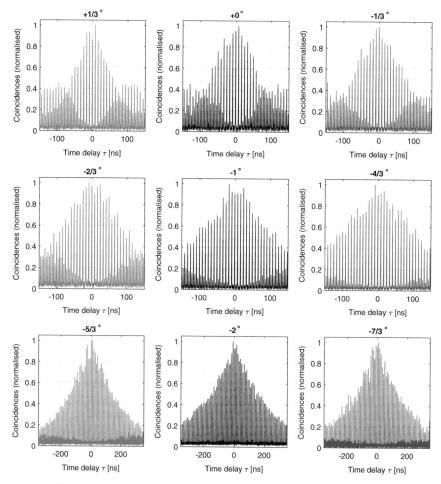

Fig. 4.2 $G_{s,i}^{(2)}(\tau)$ as a function of relative HWP angle (titles), with $0°$ the angle after initial coarse alignment. Note, the time axis in the first two rows includes ± 150 ns, while the third row is ± 350 ns in order to illustrate the effect of the HWP. Misalignment of the HWP leads to an additional peak structure caused by photon pairs leaving the SPDC cavity orthogonally polarised at odd number of round-trip differences. The position and height of the central extra peak depends on the deviation from optimum and the exact delay, respectively. As expected, the optimal angle $(-2°)$ exhibits only the main peaks separated by the effective cavity round-trip time. Getting closer to the ideal alignment (relative angle decreasing from $+1/3°$) forces the symmetric extra structure further and further away from zero time delay until it vanishes for $-2°$ (optimum) and then revives at $-7/3°$ (right lower corner)

adjacent peaks is t_{rt}, equal to two physical round trips, one traversed in each polarisation. Because of the deterministic separation of the photons on a polarising beam splitter, pairs are only detected as coincidences if their polarisation is orthogonal, in this case after an even number of physical round-trip differences. As pointed out before in Sect. 3.1.3, the HWP at 45° effectively halves the FSR of the OPO while keeping the linewidth constant, and additionally ensures double resonance of signal and idler photons by cancelling out the birefringence from the nonlinear crystal.

Slight misalignment of the HWP results in a certain probability to detect photon pairs at odd number of physical round-trip differences, leading to additional peaks halfway between the ideal structure visible in Fig. 4.2. These peaks reach their maximum when the round-trip difference of the two photons is a multiple of $\lfloor {}^{45°}/_{\Delta\alpha} \rfloor$, with $\Delta\alpha$ the deviation of the HWP angle from optimum. We can understand this as follows: the small discrepancy $\Delta\alpha$ accumulates each round-trip, creating a probability for the photons to follow orthogonal paths through the beam displacer behind the cavity and being detected as a coincidence. This probability reaches its maximum when the deviation leads to a perfectly orthogonal photon pair at an odd number of extra round-trips, where the photon delay is $\tau = {}^{(45° \times t_{rt})}/_{(2\Delta\alpha)}$. This feature allows us to set the FSR to two different values, ∼121 and ∼242 MHz, dependent on the angle of the HWP. However, in the latter case, the HWP does not affect the polarisation of the photons any more and hence does not cancel out birefringence from the nonlinear crystal. Figure 4.2 clearly shows the effect of the HWP alignment on $G_{s,i}^{(2)}(\tau)$: the closer we get to the optimal angle, the bigger the spacing becomes between zero and the maximum delay of the additional peaks. The corresponding decrease of the height of the maximum is related to the cavity decay time and subsequent a lower probability of finding coincidences with arrival times far apart from each other. In order to be able to observe the additional structure when very close to optimum, we increase the presented data time window from 300 ns to 700 ns, as displayed in the third row of Fig. 4.2. The angular step size between the individual measurements is chosen to efficiently illustrate the effect while still being reliable, with $1/3°$ corresponding to a 90° rotation of the fine adjustment screw for the HWP.

4.1.1.2 Cross-Correlation Measurement

In the case of perfect alignment of the HWP, we can use the cross-correlation function to experimentally determine the parameters of our OPO. We measure the $G_{s,i}^{(2)}(\tau)$ by recording the coincidences between the two detectors in a time window of ± 1 μs over a period of 11 min, accumulating 312000 coincidences shown in Fig. 4.3. This is sufficient to obtain statistically meaningful results. The experimental data is depicted in red (crosses) together with the fitted model following Eq. 2.77 in black. We choose the theoretical representation in the time domain [1], as the parameter space for fitting is significantly smaller. This is mainly related to the finite jitter of our detectors: in order to use $G_{s,i}^{(2)}(\tau)$ in terms of frequency modes (Eq. 2.77), the number of modes needs to be known. However, for a fit to the experimental data, this is not sufficient. The width of the individual peaks depends on a convolution of the detection time jitter with the number of frequency modes and, in our case, is dominated by the former. Therefore, accurate theoretical modelling of the experimental data requires either an artificial adjustment of the "effective" frequency mode number for Eq. 2.76 or the "effective" resolution of the detectors in Eq. 2.77. Whereas this demands the introduction of the mode number as a fitting parameter in the frequency domain, dramatically increasing the parameter space, the temporal mode representation already includes an explicit factor for the effective resolution of the detectors (ΔT).

All plots in Fig. 4.3 illustrate the same data for the measured $G_{s,i}^{(2)}(\tau)$ on three different time scales (a-c) and with two different bin sizes (c,d). The first two plots, (a) and (b), resolve the detailed peak structure arising from the multi-mode output and show the high overlap of the experimental data and the theoretical model, with a reduced $\chi^2 = 1.0062$, indicating an excellent match between theoretical model and experimental data within the error variance. The latter two sub-figures demonstrate the overall double exponential decay of the coincidence counts, expected from the cavity output, in the individual peak resolving and non-resolving regimes. It is important to point out that Fig. 4.3d is obtained by a larger time bin size, not by single-mode operation of the source as shown in Fig. 2.15. For the further evaluation of the data, we associate negative arrival time differences with the idler photon (signal arriving before idler) and positive τ with the signal photon. The HWP birefringence compensation leads to the same values for all parameters for signal and idler, supported by the experimental data within error bars. We therefore neglect subscripts, e.g. $\gamma_S = \gamma_I = \gamma$, in the following description of the classical OPO characteristics.

4.1.1.3 OPO Parameters

The double exponential decay $\exp(-2\pi\gamma|\tau|)$ of the cross-correlation function in Fig. 4.3d exhibits a cavity damping rate of $\gamma = 666$ kHz, slightly below the expected value from the mirror reflectivities and cavity length measurements. The intensity cross-correlation function follows the convolution of two comb-structured photon wave packets, each consisting of Lorentzian lines separated by the FSR and centred around the corresponding frequencies for signal and idler. Due to the condition on the signal and idler photons to be resonant simultaneously, the single photon linewidths are less than the cavity decay rate obtained from $G_{s,i}^{(2)}(\tau)$, and the FWHM linewidth simplifies to $\Delta\nu = \sqrt{\sqrt{2}-1} \times \gamma$, in our case [2, 3]. Hence, the experimentally obtained single photon linewidth for signal and idler is $\Delta\nu_{SP} = 429$ kHz, the narrowest photons to date from an SPDC-based source and well-suited to be included in atomic memory schemes using rubidium. The cavity round-trip time, illustrated in Fig. 4.3a–c as the time difference between adjacent peaks, is $t_{rt} = 8.28$ ns, corresponding to a FSR of $\nu_{FSR} = 120.77$ MHz. This is in good agreement with the values from the cavity length (Sect. 3.1.3) and the continuous-wave measurement (CW, Sect. 3.4.2) for our compensation method. The effective detector resolution, fitted to have a FWHM of $\Delta T = 930$ ps, matches the expected value from the jitter of both detectors ($2 * (425 \pm 75)$ ps) well. For comparison, the theoretical FWHM

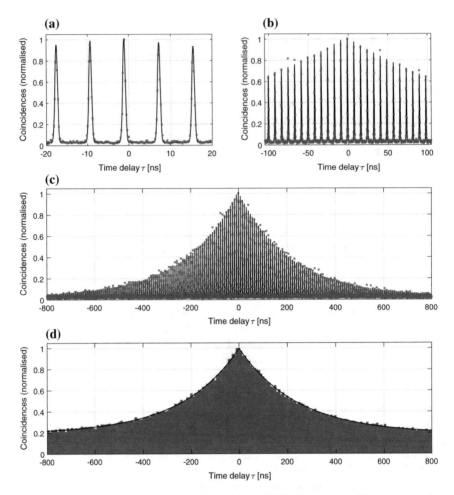

Fig. 4.3 Intensity cross-correlation function for perfect HWP alignment on different time scales and bin sizes (bs): (**a**) ±20 ns, 200.2 ps bs; (**b**) ±105 ns, 200.2 ps bs; (**c**) ±800 ns, 200.2 ps bs; (**d**) ±800 ns, 8.28 ns bs;. Data in red, theoretical model(s) in black. The overall count number is 312000 within a ±1 μs time window with an integration time of 11 min. (**a**) and (**b**) highlight the good agreement of the theoretical model and the experimental data for the central 5 and 25 peaks, respectively, while (**c**) and (**d**) illustrate the overall double exponential decay of the peaks emerging from the cavity. (**a**)–(**c**) utilise Eq. 2.77 as a fitting model, while (**d**) follows a simpler double exponential decay given by $\exp(-2\pi\gamma|\tau|)$

Table 4.1 OPO output parameters for different measurement methods including uncertainties

Parameter	Single photon $G_{s,i}^{(2)}(\tau)$	CW cavity characterisation	Theory (specifications)
Cavity decay rate γ [kHz]	666 (16)	750 (150)	<760
Single photon linewidth $\Delta\nu_{SP}$ [kHz]	429 (10)	480 (100)	<490
Round-trip time t_{rt} [ns]	8.280 (0.005)	8.27 (0.07)	8.27
FSR ν_{FSR} [MHz]	120.77 (0.05)	121(1)	121
Finesse $\mathcal{F}$	181 (4)	160 (30)	>161
Peak width ΔT [ps]	930 (5)	–	–

of the peaks from Eq. 2.77 is $\Delta T \approx 1.9$ ps, utilising the values presented in this paragraph for all free parameters and the correct number of frequency modes (1867) in the full phase-matching envelope. This proves the dominance of the detector jitter over the frequency mode contribution to the peak width in multi-mode operation of the source. Table 4.1 gives an overview of the parameters derived from the single photon $G_{s,i}^{(2)}(\tau)$ measurement and compares them to the former obtained values. Due to the high number of resolved peaks and the quality of the acquired data, the single photon characterisation is more than an order of magnitude more accurate than the other methods, as implied by the low uncertainties.

After establishing the basic parameters of our OPO, we can now derive more advanced characteristics of the single photons from $G_{s,i}^{(2)}(\tau)$ and compare them to theoretical predictions. First, the biphoton correlation time τ_{cor} is defined as the FWHM of the double exponential decay. The theoretical value for an OPO is given by [4]

$$\tau_{cor} = \frac{1.39}{2\pi\gamma} = 332 \text{ ns.} \tag{4.1}$$

This is in excellent agreement with the value observed in our measurement presented in Fig. 4.3d: $\tau_{coh} = (331 \pm 2)$ ns. We further define the coherence time τ_{coh} of the source, directly related to the spatial length l_{coh} of the single photons via the speed of light c. For a Lorentzian frequency mode, as expected from the output of an optical resonator, the coherence time and length are given as [5]

$$\tau_{coh} = \frac{1}{\pi\Delta\nu_{SP}}, \tag{4.2}$$

$$l_{coh} = \tau_{coh}c. \tag{4.3}$$

For our source, Eqs. 4.2 and 4.3 lead to $\tau_{coh} = (740 \pm 20)$ ns and $l_{coh} = (222 \pm 5)$ m for coherence time and length, respectively. These are the longest photons from SPDC to date, offering new possibilities in quantum optics experiments, e.g. for loophole-free quantum causality (switch) experiments [6, 7]. In Sect. 4.1.2, we are using the coherence time to define a symmetric time window with a width of $2\tau_{coh}$ in order to determine the spectral brightness of the source.

4.1.1.4 Post-processing Filter

The comb-like structure of the coincidence measurements shown in Fig. 4.3, arising from the multi-mode operation of the source, is a reoccurring feature in all measurements presented in this chapter. In order to derive some of the upcoming results, we implement a post-selection filtering scheme in the evaluation code of the data: as the arrival time differences of the photons are very well known, we only select the measured data in a smaller time window of width $\tau_f = 1.07$ ns (13 time bins with 82.3 ps width each) around each of the peaks within τ_{coin}. We therefore eliminate coincidences which are not single photon pairs from the OPO in a process similar to gating the detectors. Figure 4.4 shows an example of the filtering process with $\tau_{coin} = \pm 20$ ns for the results in this section. The symmetrical coincidence time window (green) includes five peaks of the $G^{(2)}_{s,i}(\tau)$ measurement, each filtered by the code in post-processing (blue), leading to an effective coincidence time window of 5.35 ns in this example. In the upcoming sections of this chapter, the implementation of this filtering step on the experimental data will be explicitly indicated each time together with the resulting effective coincidence window.

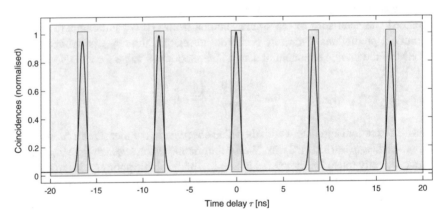

Fig. 4.4 Post-selection filtering (gating) scheme. Cross-correlation function in black, total coincidence window (τ_{coin}) in green, filtering window (τ_f) around each peak in blue. First, the large coincidence window is implemented to find all detected photon pairs within τ_{coin}, chosen dependent on the application. Then, the filtering step is performed to obtain the counts inside the multiple small filtering windows, each with a width of $\tau_f = 1.07$ ns, leading to an effective time window of 5.35 ns in this example

Table 4.2 Background corrected spectral brightness (SB) in a time window of 1500 ns, filtered and unfiltered

Heralding rate [kHz]			27	37	48	63	117
SB	$\frac{\text{photon pairs / s}}{\text{mW*MHz}}$	(unfiltered)	4840 (380)	4570 (360)	4240 (320)	3850 (300)	4700 (300)
SB	$\frac{\text{photon pairs / s}}{\text{mW*MHz}}$	(filtered)	4350 (340)	3950 (310)	3760 (290)	3360 (280)	4080 (300)

4.1.2 Spectral Brightness

The spectral brightness of the single photon source is determined by calculating the number of coincidences per second in a symmetric time window $\tau_{coin} = 1500$ ns (double the coherence time, $2\tau_{coh}$) at multiple heralding rates, associated with individual pump powers. In order to obtain correct and meaningful results, the measured count rate needs to be adjusted for unavoidable accidental coincidences, depending on the single photon rates and the detection window [8]

$$R_{ac} = R_S R_I \tau_{coin}. \tag{4.4}$$

R_{ac} is calculated individually for each data set and subtracted from the measured coincidence rate. Due to the big time window, almost two orders of magnitude higher than in experiments with broadband sources, applied to derive the spectral brightness, the effect of the accidental rate is significant. The experimental data is analysed in two separate ways: in the unfiltered case, presented in the upper row in Table 4.2, all detected coincidences in the full time window are summed. The results in the lower row, however, are obtained by the filtering process described in Sect. 4.1.1. Furthermore, all spectral brightness values are only corrected for the quantum efficiency of the detectors (56%) at the single photon wavelength and can therefore be additionally improved by increasing the fibre coupling, escape efficiency of the cavity or overall transmission of all extra optical elements between the cavity and the fibre couplers. The directly measured coincidence rate is multiplied with the creation probability of the central degenerate frequency mode (0.22%), meaning that the presented results describe the single-mode brightness, expected after extra spectral filtering by e.g. the mode-cleaning cavity.

The directly measured result for the spectral brightness of our source in multimode operation, only accounting for background, is $B = 1.13 \times 10^6 \frac{\text{photon pairs / s}}{\text{mW*MHz}}$. The adjusted values in Table 4.2, additionally correcting for the detector efficiency and the creation probability of the degenerate mode, lead to a mean of $B_f = 3900 \pm 370 \frac{\text{photon pairs / s}}{\text{mW*MHz}}$ and $B_u = 4440 \pm 400 \frac{\text{photon pairs / s}}{\text{mW*MHz}}$ for the filtered and unfiltered case, respectively, matching within their uncertainties. This is the brightest source of its kind to date, outperforming the former leading value from Ref. [9] by a factor of three. High spectral brightness is essential for applications of the source in quantum communication and other photonic quantum technologies [10].

4.2 Quantum Characterisation

4.2.1 Multi-photon Suppression

The quantum statistics of our single photon source can be analysed by measuring the intensity auto-correlation function $g_{s,s}^{(2)}(\tau)$ and more importantly the special case of its value at zero time delay $g_{s,s}^{(2)}(0)$, first experimentally demonstrated in the quantum regime by Grangier and coworkers [11]. As already discussed in Sect. 2.4.3, $\left(1 - g_{s,s}^{(2)}(0)\right)$ is a measure of the multi-photon suppression of the single photon source, with a value <1 indicating non-classical behaviour and $g_{s,s}^{(2)}(0) = 0$ for a perfect single photon Fock state without higher order contributions. The auto-correlation measurement is realised experimentally in a Hanbury Brown and Twiss setup [12] (see Fig. 2.16) added in fibre to the experimental setup (Fig. 3.22) at the single photon connectors: we herald on one single photon (idler) while splitting the other (signal) probabilistically on a 50/50 beam splitter and subsequently monitor coincidences between all combinations of the individual channels. The formula used to calculate the presented results in this section is given similar to Eq. 2.92:

$$g_{s,s}^{(2)}(\tau) = \frac{N_{ssi}(\tau)N_0}{N_{si}(0)N_{si}(\tau)}, \tag{4.5}$$

with N_{si} (signal-idler) and N_{ssi} (signal-signal-idler, indicating the undesired creation of more than one photon pair) the double and triple coincidences between individual detectors and N_0 the amount of trigger (heralding) events recorded in the idler arm. N_{si} and N_{ssi} are determined by measuring $G_{s,i}^{(2)}(\tau)$ between the corresponding channels and summing over all double and triple coincidences in a (post-filtered) time window, respectively. Additionally, we multiply N_{ssi} with a factor of two to account for possible bunching, as customary in the field [3, 13].[1] $g_{s,s}^{(2)}(\tau)$ is actually defined for a precise τ (Eq. 4.5) but can only be determined experimentally by using finite time windows, implemented in post-processing. Accordingly, the presented results in this section have to be understood as upper limits, smeared out by the time window. All error analysis on the data is performed by taking the square root of the number of counts, with the amount of triple coincidence in the measurement window for $g_{s,s}^{(2)}(\tau)$ being the dominating source of uncertainty. Other possible error sources, including dark counts and dead times of the detectors or stray light contributions, are not incorporated into the error analysis in this section.

An ideal single photon source never emits more than one signal photon at a time, a process called anti-bunching, and therefore the expected value for triple coincidences in Eq. 4.5 is zero, leading to $g_{s,s}^{(2)}(0) = 0$. However, in an actual experiment, the achievable minimal value for $g_{s,s}^{(2)}(0)$ depends on a variety of parameters used to obtain and process the data: the width τ_f of the implemented post-filtering window

---

[1]Unfortunately, private communication with the authors of Ref. [13] could not fully explain the origin of this factor.

around each peak (see Sect. 2.4.3), the overall width of the coincidence window τ_{coin} (= $2\tau_c$, see Sect. 2.4.3), symmetric around zero delay, and the heralding rate R_0. The upcoming paragraphs analyse the dependency of $g_{s,s}^{(2)}(0)$ on all three parameters independently and compare their values to theoretical predictions. The section concludes with the presentation of the time dependent intensity auto-correlation function $g_{s,s}^{(2)}(\tau)$ and its dependence on the coincidence window. None of the upcoming results are background corrected which would decrease the values for $g_{s,s}^{(2)}(0)$ even further.

4.2.1.1 Filtering Window

We first investigate the dependence of $g_{s,s}^{(2)}(0)$ on the implemented filtering window τ_f (described in Fig. 4.4). Changing τ_f effectively acts as a gating operation to post-select time intervals where the single photons are expected to arrive at the individual detectors, defined by multiples of the round-trip time of the cavity, similar to Sect. 4.1.2 and Fig. 4.4. Due to the multimode operation of the source in the presented experiments, these time delays between coincidences are very well known and only broadened by the detector jitter as derived in Sect. 4.1.1. The ratio of signal to accidental counts decreases for bigger filtering windows, as more coincidences outside the expected arrival time differences ("accidents") are included in the calculations. According to Eq. 2.93 (Sect. 2.4.3), $g_{s,s}^{(2)}(0)$ is meant to be linearly increasing as a function of τ_f, if $R_0 \tau_f P \ll 1$, with P the number of included peaks. This behaviour is illustrated in Fig. 4.5. Here, all the values for the auto-correlation function are calculated from $N_0 \approx 200 * 10^6$ idler events with a heralding rate of $R_0 = 16\,\text{kHz}$ and a coincidence window of 250 ns, approximately the FWHM of the exponential decay in Fig. 4.3d. The combination of τ_f and τ_{coin} for Fig. 4.5 results in effective time windows between $12.8 - 130$ ns, with a finally chosen value of 33.2 ns. Filtering windows up to $\tau_f = 1.73$ ns are below the width of the peaks, hence mainly capturing coincidences from photons created by the OPO. Above this value, additional coincidences and especially triple events are more and more dominated by accidentals, further increasing $g_{s,s}^{(2)}(0)$ due to the large coincidence time window necessary for our source. We choose a value of 1.07 ns (= 13 time bins) around each individual peak maximum, well within the borders of the individual coincidence peak to avoid events clearly not originating from the OPO (red data point in Fig. 4.5). The filtering window covers the full width at 30% of the maximal peak height, leading to a high ratio of signal to accidental coincidences.

4.2.1.2 Coincidence Window

After selecting a filter window of $\tau_f = 1.07$ ns, incorporating most of the photons actually emitted by the source, we analyse the experimental data to study the dependence of $g_{s,s}^{(2)}(0)$ on the coincidence window τ_{coin}. In experiments with broadband photons (e.g. from SPDC or quantum dots), the coincidence window is typically

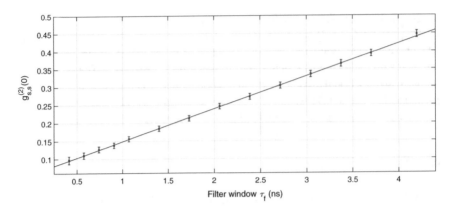

Fig. 4.5 Intensity auto-correlation at zero time delay ($g_{s,s}^{(2)}(0)$) as a function of the filtering window width τ_f. The linear growth is in good agreement with the theoretical predictions from Eq. 2.93, fitted in black. The red data point ($\tau_f = 1.07$ ns) is the width used in further experiments, chosen for a good signal to accidental ratio and small uncertainties. The value corresponds to the full width at approximately 30% of the individual peak heights shown e.g. in Fig. 4.3a. The overall coincidence and corresponding effective time windows are 250 ns and 33.2 ns, respectively

chosen on the order of the coherence time ($\sim$ns), leading to a rate of accidental triple coincidences which is 5–6 orders of magnitude below the values expected for our source. In order to make meaningful statements on the auto-correlation measurement in the field of narrow-band single photon sources, it is common practise to use windows far below the extension of the photons ($\tau_{coin} \ll 2\tau_{coh}$), either by using smaller periods of time in the first place [3] or extrapolating down from larger windows [14]. Here, we analyse the data for a variety of τ_{coin} between 50 and 1500 ns, corresponding to effective time spans of $7.5 - 194$ ns, with the maximal coincidence time chosen to be twice the coherence time. The overall number of trigger events is again $N_0 \approx 200 * 10^6$ at a heralding rate of $R_0 = 16$ kHz. The experimental results are shown in Fig. 4.6. Similar to the dependency of $g_{s,s}^{(2)}(0)$ on τ_f and, again, in good agreement with Eq. 2.93, the calculations exhibit a linear rise of the auto-correlation values governed by the decreasing signal to accidental events ratio at higher τ_{coin}. However, in this case, the lower ratio is caused by the cavity decay, resulting in a smaller probability of detecting single photon pairs at vastly different arrival times while the background stays constant, see Sect. 4.1.1. As this ratio is generally still high compared to using large values for τ_f, the slopes of $g_{s,s}^{(2)}(0)$ depends on the effective time windows for Figs. 4.5 and 4.6, which differ by a factor of $\sim$1.6. This discrepancy arises solely due to different contributions of the background to the evaluated data. For all further calculations in this section we choose $\tau_{coin} = 250$ ns, close to the correlation time of the photons. This allows to accumulate sufficient coincidence counts for meaningful statistics in a reasonable time frame, without being dominated by background noise.

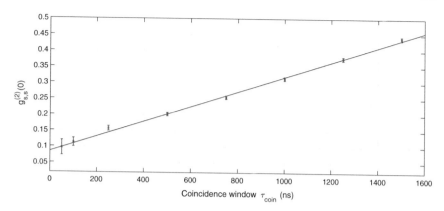

Fig. 4.6 Intensity auto-correlation at zero time delay ($g_{s,s}^{(2)}(0)$) as a function of the coincidence window τ_{coin}. Similar to Fig. 4.5, we can observe a linear increase shown in black (see Sect. 2.4.3) of $g_{s,s}^{(2)}(0)$, arising from the declining signal to accidental ratio of triple coincidences. The red data point at $\tau_{coin} = 250$ ns corresponds to the coincidence time window used in further experiments, approximately the FWHM of the exponential decaying coincidences (e.g. Fig. 4.3d). Again, the resulting effective time window is 33.2 ns at this point

4.2.1.3 Heralding Rate

The last parameter investigated for its effect on $g_{s,s}^{(2)}(0)$ is the heralding rate $R_0 = N_0/t_m$, where t_m is the total measurement time for the obtained data point. In Sect. 2.4.3, we derive the theoretical dependence according to [13]:

$$g_{s,s}^{(2)}(0) = 2 - \frac{2\,(R_0 \Delta\nu_{SP})^2}{\left[R_0^2 + R_0 \Delta\nu_{SP}\right]^2}, \tag{4.6}$$

with $\Delta\nu_{SP}$ the single photon linewidth. This is the only free variable in Eq. 4.6 which can therefore be used to independently validate the linewidth calculations from the cross-correlation measurements. We can see that $g_{s,s}^{(2)}(0)$ approaches zero for very small heralding rates. This is expected as the probability for multi-photon emission in SPDC sources depends quadratically on the intensity of the incoming pump field, while the single photon events increase linearly.

The results for multiple heralding rates are presented in Fig. 4.7. Every data point consists of $N_0 \approx 200 * 10^6$ triggering events, leading to total measurement times between 15 min ($R_0 = 225$ kHz) and 16 hours ($R_0 = 3.5$ kHz). The implemented filtering time is 1.07 ns, with an overall coincidence window of 250 ns, resulting in an effective time window of 33.2 ns, as prior derived in this section. This is slightly smaller than in similar works [3, 14], but offers higher accuracy as the overall integration time is chosen to collect sufficient statistics for reasonably small error bars. It is important to point out that all the results obtained are not background corrected, which would improve $g_{s,s}^{(2)}(0)$ even further. The lowest value of $g_{s,s}^{(2)}(0) = 0.032 \pm 0.003, 277$

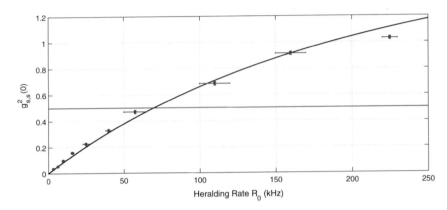

Fig. 4.7 Intensity auto-correlation at zero time delay $(g_{s,s}^{(2)}(0))$ as a function of the heralding rate R_0. Experimental data in blue, theoretical model in black, $g_{s,s}^{(2)}(0) = 0.5$ in red. The fitted curve (Eq. 4.6) independently confirms the previously calculated linewidth in Sect. 4.1.1: here, $\Delta\nu_{SP} = 450 \pm 32$ kHz. The larger uncertainties on some of the heralding rates arise from temperature fluctuations over the measurement period, changing on a daily basis

standard deviations below the classical limit of $g_{s,s}^{(2)}(0) = 1$, is obtained at a heralding rate $R_0 = 3.5$ kHz and clearly shows the single photon nature of the source. The theoretical model (Eq. 4.6) in Fig. 4.7 overlaps well with the experimental behaviour. The fit determines the only free parameter to be $\Delta\nu_{SP} = 450 \pm 32$ kHz, separately confirming the linewidth calculations from the cross-correlation measurements in Sect. 4.1.1. Additionally, the results exhibit anti-bunching below 0.5 [15] up to heralding rates of $R_0 = 70$ kHz, depicted in red in Fig. 4.7. The slightly larger error bars on some of the heralding rates mainly arise from instabilities of the triple resonance condition within the measurement period, affecting the measured rate.

4.2.1.4 Auto-Correlation Function $g_{s,s}^{(2)}(\tau)$ at $\tau \neq 0$

Finally, after thorough analysis of the behaviour of the auto-correlation function at zero time delay, the experimental data at a heralding rate of $R_0 = 16$ kHz is examined for its time dependence to acquire $g_{s,s}^{(2)}(\tau)$. Although $g_{s,s}^{(2)}(0)$ is the more important figure of merit, $g_{s,s}^{(2)}(\tau)$ is still very useful to demonstrate the expected behaviour and validate the assumptions made throughout this section. In Sect. 2.4.3, we derive the theoretical framework for the time dependent auto-correlation function and observed some interesting features: the minimum value of $g_{s,s}^{(2)}(\tau)$ depends on the heralding rate as well as the coincidence window and is meant to be constant for a certain amount of time delay around zero, whereas $g_{s,s}^{(2)}(\tau) = 1$ for large τ, independent of the type of light source. In the transition region, the dependence on τ_{coin} can only be calculated numerically, but approximately follows a Gaussian function $\propto \exp\left[-(w\tau)^2\right]$ [16], with w a constant incorporating the coincidence window and single photon linewidth.

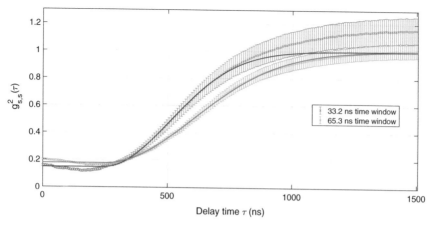

Fig. 4.8 Time-dependent intensity auto-correlation function $g_{s,s}^{(2)}(\tau)$ for two effective time windows, 33.2 ns (blue) and 65.3 ns (green). The data and the theoretical fit (thick lines) overlap well and exhibit the expected behaviour: (almost) constant for small time delays, Gaussian-like rise with time window dependent slope causing an intersection in the transition region and, finally, $g_{s,s}^{(2)}(\tau) = 1$ for large τ

As the exact starting position and width of the transition region differ for individual τ_{coin}, the traces are expected to intersect.

Our results are shown in Fig. 4.8 for two different effective time windows, 33.2 ns (31 bins) and 65.3 ns (61 bins), correspond to a coincidence window of 250 ns and 500 ns, respectively. The slight imbalance between the window ratios (1.97 and 2) arises from the total amount of peaks $m = n + 1$ in the coincidence window, with the rounded integer $n = \lfloor {}^{\tau_{coin}}/_{t_{rt}} \rfloor$ accounting for all included peaks apart from the one at zero delay, see e.g. Fig. 4.4. The higher coincidence window (green) leads to a broader and less distinct dip of $g_{s,s}^{(2)}(0) = 0.2$ compared to $g_{s,s}^{(2)}(0) = 0.17$ for the smaller window. In general, the fitted curves reproduce the experimental data within the error bars, calculated as the square root of the count numbers and hence dominated by the uncertainty on triple coincidences, fairly well: more significant deviations outside the uncertainty are mainly limited to the smaller time window (blue). The preliminary reduction of $g_{s,s}^{(2)}(\tau)$ with increasing τ for both data sets is unexpected and cannot be fully explained within the error margins of the data. The values > 1 at large delays for the small coincidence window can be understood by additionally taking detector dead time and dark counts into consideration, where especially the dead time effects the smaller time window statistics more. Overall, the experimental data and the fitted curves exhibit the expected behaviour of different $g_{s,s}^{(2)}(0)$ and dip width, including a crossing of the traces, and reaching $g_{s,s}^{(2)}(\tau) = 1$, for large time differences.

4.2.2 Indistinguishability

Indistinguishability is a key component of generating spatially entangled states for photons. The attribute indistinguishable refers to all possible degrees of freedom that are not entangled and include but are not limited to: polarisation, spatial, temporal and frequency mode. A way to quantify the indistinguishability is in a Hong-Ou-Mandel (HOM) experiment [17], exhibiting purely non-classical two photon interference. In Sect. 2.4.4, we introduce the theoretical model for the HOM effect on the temporal degree of freedom for a SPDC source [18] and furthermore the unique temporal shape of our photons in particular [19]. The results presented in this section illustrate the indistinguishability, temporal length and reoccurring non-classical interference (multi-mode operation) of the narrow-band single photons.

4.2.2.1 HOM Setup and Alignment

The schematic HOM setup is shown in Fig. 4.9. The delay line $d\tau$ in the signal arm is implemented in a combination of additional fibre for static large delays of multiples of the cavity round-trip length $l_{rt} = 2.48$ m ($\mapsto t_{rt} = 8.28$ ns) and a motorised translation stage (see Fig. 3.22) for fine adjustment of the single photon arrival times within ± 20 mm ($\mapsto \pm 67$ ps). The 50/50 beam splitter is realised in optical fibre (FBS) in order to minimise losses and complexity that arise from a free-space setup. After the beam splitter, the output ports are separately detected on SPCMs and analysed for coincidences. Unfortunately, this (nearly) complete in-fibre implementation of the HOM setup makes the matching of all the degrees of freedom of the photon pairs on the beam splitter difficult. Especially the fragile polarisation overlap, affected by temperature changes or bending of the individual optical fibres, is challenging and requires additional fibre polarisation control in both arms and regular careful alignment.

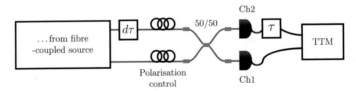

Fig. 4.9 Schematic setup to perform HOM interference experiments, with the temporal mode (arrival time dependent on $d\tau$) as the distinguishability parameter. The time delay $d\tau$ is introduced by a motorised translation stage to measure the width and depth of the HOM dip and additional fibres of various lengths to measure adjacent dips. Individual polarisation control in signal and idler arm is necessary to match both channels at the 50/50 fibre beam splitter (FBS). Photon detection is achieved by respective SPCMs and recorded by a time tagging module (TTM), introducing an additional electronic delay τ

In order to match the photons, the signal and idler paths are aligned and inspected on a daily basis. Prior to any measurement, the SHG cavity and the coupling of the pump light into the OPO is optimised for low-noise operation. It can be demanding to tune the temperature of the SPDC crystal to triple resonance, as indicated in Sect. 3.3.3. However, in case of HOM measurements, the photons do not need to be exactly at the rubidium transition. Thus the stabilisation loop to the atomic reference is not required, thereby allowing alignment of the triple resonance via an electronic offset applied to the locking point of the SPDC cavity. Tuning this extra parameter changes the resonance condition for pump and single photon linewidths differently and hence simplifies achieving and keeping the triple resonance condition.

In the next step, a custom-made fibre with a length corresponding to the desired delay of the interference dip (see the theory in Sect. 2.4.4 for more details) needs to be inserted into the signal arm. Every fibre is individually spliced to the desired length and tested for maximal obtainable transmission before being used in the experiment. Although each individual temporal mode of the single photon pairs is very long, the multi-mode output of the source results in very well defined arrival time differences, as illustrated in Fig. 4.3a–b. This means that the arrival time of the photons at the beam splitter for the HOM measurement must be matched very accurately, namely within ∼4 mm ($\mapsto$ ±13 ps), demanding high precision manufacturing of the fibre length.

After the relevant optical fibre is embedded in the experimental setup, the polarisation of the two photons at the beam splitter is arbitrary due to small birefringence from bending, temperature changes or imperfections of each fibre. Signal and idler paths both contain fibre polarisation controllers to account for this undesired rotation. We optimise the polarisation by adding a fibre-based polarisation beam splitter (FPBS) to one of the output arms of the FBS and monitor the splitting on two detectors. First, the signal arm is blocked and the counts on one of the FPBS outputs are minimised to background level (maximising the other output). This is verified by blocking all incoming light and determining if there is any residual change in the single photon count rate. Next, the same procedure is repeated but with the idler instead of the signal arm blocked. The exact polarisation of both photons at the BS after this procedure is still unknown, but due to the common path of propagation from the FBS towards and inside the FPBS, signal and idler now have the same polarisation at the designated point of interference. As we introduce the distinguishability on a different degree of freedom (arrival time), this knowledge about the polarisation state is sufficient for further measurements.

4.2.2.2 Measurement Settings and Data Acquisition

The experimental data used to obtain the characteristic HOM dip(s) is acquired by looking at the cross-correlation function $G^{(2)}_{s,i}(\tau)$ between the two output ports of the fibre-based BS (mixture of signal and idler). However, processing the data is not trivial and requires an extension to the evaluation code written in Matlab. In the case of HOM interference, the cross-correlation function not only depends on the time

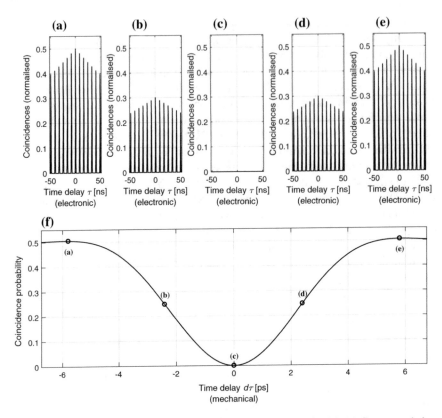

Fig. 4.10 Theoretical HOM interference behaviour for the central dip. (**a**)–(**e**) Cross-correlation function $G^{(2)}_{s,i}(\tau)$ between the two output ports of the BS, dependent on the electronically-introduced τ from the TTM, for different time delays $d\tau = [-5.8, -2.4, 0, 2.4, 5.8]$ ps. (**f**) Expected HOM interference dip as a function of $d\tau$, mechanically introduced by mirrors mounted on a motorised translation stage. The dots are the calculated coincidence probabilities from the cross-correlation functions above (**a**)–(**e**) in the according position (left to right). The closer we get to perfectly matched temporal delays of the photons, the smaller the peak structure of $G^{(2)}_{s,i}(\tau)$ becomes until it vanishes completely for ideal overlap (**c**)

delay τ, electronically introduced by the TTM with a resolution of 82.3 ps, but also on an extra delay $d\tau$, mechanically added via an extra optical fibre and a pair of mirrors mounted on a motorised translation stage, and adjustable on a micrometer scale ($\mapsto$ sub-picosecond photon travel time). The actual HOM dip is a function of $d\tau$, but each data point consists of a full measurement and post-processing of $G^{(2)}_{s,i}(\tau)$ for each individual $d\tau$.

The data acquisition is illustrated in Fig. 4.10. Changes in $d\tau$ can affect $G^{(2)}_{s,i}(\tau)$ in three different ways: if the mechanical time delay does not match the possible arrival times of the photons on the beam splitter, no HOM interference is observable, however, the coincidence probability is halved as shown in Fig. 4.10a, e, f. The comb-like structure shows the same decay, peak separation and peak width as derived and analysed in Sect. 4.1.1. The additional time delay introduced by the motorised translation stage, excluding the extra fibre for now, is orders of magnitude below the resolvable level of the detectors and therefore does not shift $G_{s,i}^{(2)}(\tau)$ within the measurement accuracy.

In the second case, presented in Fig. 4.10b, d, f, the biphoton wave packets partially overlap. This leads to an overall decay in the peak height of the observed coincidences for the full extend of the photons, as the probability of both photons travelling to the same detector increases. In other words, matching the arrival times of the photons on the picosecond scale results in non-classical interference detectable at 100s of nanoseconds delay. This is quite remarkable and, with some considerations, straight forward to understand: each photon has a certain exponentially decaying probability of leaving the OPO at integer multiples of the round trip times (t_{rt}) which stretches over the whole coherence length ($\sim$750 ns). Accordingly, interference is measurable on that time scale when one photon path length is properly matched with its partner, reducing the probability to detect coincidences. This concept is another independent proof of the long coherence and narrow linewidth of the generated photon pairs.

The last scenario, shown in Fig. 4.10c, f, explains the case of perfectly matched temporal modes. All arriving photons interfere and move pairwise along either channel towards the detectors. The peak structure in $G_{s,i}^{(2)}(\tau)$ vanishes completely as there are no correlated coincidences measurable within τ_{coin}, leading to the minimum of the dip in the coincidence probability. However, in a realistic experiment there will always be some residual coincidences arising from background events, dark counts or a small mismatch in any other degree of freedom, e.g. the polarisation mode, that creates deviations from the ideal case. The visibility of the HOM interference is defined as the difference between the temporally unmatched (green) and matched (red) case, divided by the sum of the two (Eq. 2.115).

Although described above for no time delay, the explanation of the three different regimes of biphoton overlap can be mapped directly onto the scenario where an additional optical fibre introduces a large time delay $d\tau = n \times t_{rt}/2$, with $n \in \mathbb{N}$. In the multi-mode operation case of the source described here, this extra optical fibre can be implemented in the path of the signal photon in order to observe revivals of the HOM dip, derived in Sect. 2.4.4. Again, fine tuning of the arrival time difference is achieved with the motorised translation stage, allowing to scan past the dip and to correct for possible imperfections of the fibre length. Due to the underlying nature of the interference in the multi-mode case, the FWFM of the HOM dips is only dependent on the phase matching envelope of the nonlinear crystal and, hence, stays the same for all dips. The maximal achievable visibility, on the other hand, depends on the changing indistinguishability of the biphoton amplitude and slightly decreases with growing length of the delay line.

We perform the HOM measurements at equal arrival time and at a variety of different optical fibre delays of length $\sim$ [2.5, 5, 10, 100, 105] meters equivalent to one, two, four, 40 and 42 effective cavity round-trips, respectively. Each interference fringe is measured multiple times to eliminate systematic errors and determine the reproducibility of the effect. The results are shown in Fig. 4.11. Every data point

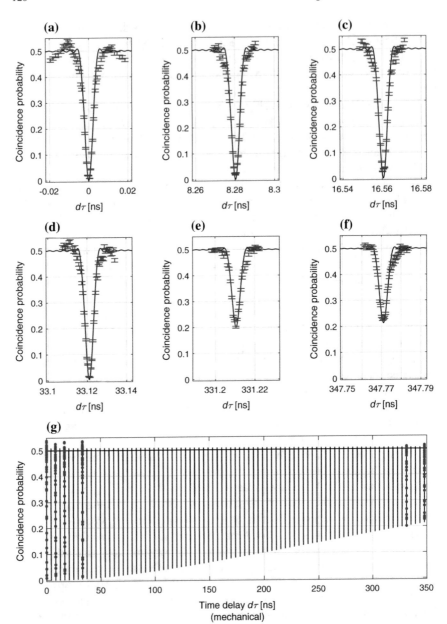

Fig. 4.11 HOM dip revivals for different time delays $d\tau$ introduced by optical fibres in the path of the signal photon. Theoretical model in black, experimental data in blue. (**a**)–(**f**) Individual dips around $d\tau = [0, 2, 4, 8, 80, 84] \times {}^{t_{rt}}/2$ ([0, 2.5, 5, 10, 100, 105] meter physical delay) at high resolution. (**g**) Entire structure, demonstrating the measured delays. The data is excellently described by the theoretical model. Small imperfections in the achievable minimum and subsequently the visibility are likely to result from unaccounted mismatch in other degrees of freedom and dispersion in the optical fibre. The small deviations in coincidence probability when approaching 0.5 can be explained by changes in the fibre coupling efficiency

corresponds to a measurement of $G_{s,i}^{(2)}(\tau)$ for around $7-10$ seconds at a single photon rate of ~ 40 kHz, leading to total numbers $N_1 \approx N_2$ between $(3-4) \times 10^5$ counts on both detectors. The step size for $d\tau$ is chosen between $0.5-1.3$ ps $(150-400\,\mu\text{m})$, with the higher resolution at the steep slopes of the dips. The coincidence, filtering and effective time windows are selected to be 1000 ns, 1.07 ns and 129 ns, respectively, sufficiently large to accumulate enough statistics in a short period of time. Excessive data analysis shows that the interference visibility (Eq. 2.115), the figure of merit in these measurements, is largely independent of the implemented time windows within the uncertainty in our experiments. In order to eliminate effects of varying single photon rates on the number of coincidences, we sum all two-photon events N_{12} in τ_{coin} and divide by the total number of single counts to obtain P_c, the coincidence probability:

$$P_c = \frac{N_{12}}{N_1 + N_2}, \tag{4.7}$$

with N_{12}, N_1 and N_2 all background corrected. Equation 4.7 can be interpreted as a kind of heralding rate and is further normalised to $P_c = 0.5$ far outside the dips. The error bars of the individual data points are calculated by taking the square root of the coincidences in the time window, as this is the dominating part of the uncertainty. Additional sources of error, e.g. mismatch in any other degree of freedom or losses and dispersion inside the optical fibre, are difficult to estimate and are therefore excluded in the error analysis. However, they are very likely to account for the mismatch between the predicted and measured visibility.

4.2.2.3 Results

The experimental data (blue) and theoretical model (black) in Fig. 4.11a–f exhibit excellent agreement. It is important to point out that the theory is not fitted to the data. All free parameters like cavity linewidth or FSR, are calculated from the measurements presented in this chapter and derived from Eq. 2.114. The average measured full width at full maximum of the dips is 14.0 ± 0.9 ps. The maximally achieved visibility (at zero time delay) is (96.7 ± 3.4) % for $\tau_{coin} = 1000$ ns and $\tau_f = 1.07$ ns. Decreasing the coincidence time window improves the signal to accidental coincidence ratio, leading to higher visibilities up to $V = (98.5 \pm 4.5)$ % at $\tau_{coin} = 250$ ns. Smaller windows are possible, but have larger associated error bars due to the lower amount of events included in the calculation.

The theoretically predicted and experimentally achieved visibilities are presented in Table 4.3 and Fig. 4.12. As expected, the experimentally measured values are generally lower than the predictions due to scattered background photons, causing accidental coincidences, and the all-fibre nature of the experiment, making it challenging to achieve ultra-high indistinguishability due to dispersion. The former is still a problem in our measurements, although the implementation of the filtering time window in post-processing acts similar to a gating operation on the detectors. Excluding this post-selecting step leads to visibilities $\sim 10\%$ lower than the values

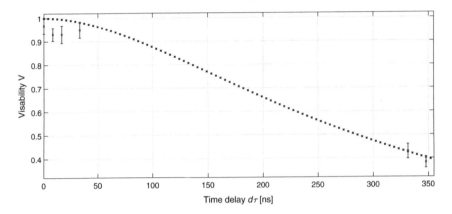

Fig. 4.12 Theoretical (black) and experimental (blue) HOM visibilities for different dip revivals around time delays $d\tau = n \times {}^{t_{rt}}/_2, n \in \mathbb{N}$ (corresponding dips illustrated in Fig. 4.11g). The lower than expected values in the experimental data arise from additional distinguishability in different degrees of freedom and accidental background events that are not fully accounted for

Table 4.3 Theoretical, experimental and relative HOM interference visibility in percent. Uncertainties describe the error bars on the lowest measured point

Peak number $n = \lfloor {}^{2d\tau}/_{t_{rt}} \rfloor$	Visibility V_{th} (theory)	Visibility V_{exp} (experiment)	Relative visibility $V_{rel} = {}^{V_{exp}}/_{V_{th}}$
0	100	96.7 ± 3.4	96.7 ± 3.4
2	99.9	93.0 ± 2.7	93.1 ± 2.7
4	99.5	90.0 ± 3.7	90.5 ± 3.7
8	98.3	94.9 ± 3.4	96.5 ± 3.5
80	42.5	42.8 ± 3.1	100.7 ± 7.3
84	40.1	38.2 ± 2.4	95.3 ± 6.0

stated in Table 4.3. The presented data is, to our knowledge, the first demonstration of HOM interference (and dip revivals) of photons arriving at time delays of 105 m (corresponding to ∼ 70 m of optical fibre in the experiment), further proving the exceptional coherence length of the photons. This can be very useful, e.g. for measurements on quantum foundations like quantum causality.

References

1. Herzog, U., Scholz, M., Benson, O.: Theory of biphoton generation in a single-resonant optical parametric oscillator far below threshold. Phys. Rev. A **77**, 023826 (2008)
2. Scholz, M., Koch, L., Benson, O.: Analytical treatment of spectral properties and signal-idler intensity correlations for a double-resonant optical parametric oscillator far below threshold. Optics Commun. **282**, 3518–3523 (2009)

3. Scholz, M., Koch, L., Benson, O.: Statistics of narrow-band single photons for quantum memories generated by ultrabright cavity-enhanced parametric down-conversion. Phys. Rev. Lett. **102**, 063603 (2009)
4. Lu, Y.J., Ou, Z.Y.: Optical parametric oscillator far below threshold: experiment versus theory. Phys. Rev. A **62**, 033804 (2000)
5. Saleh, B., Teich, M.: Fundamentals of Photonics. Wiley Series in Pure and Applied Optics, 2nd edn. Wiley (2007)
6. Chiribella, G., D'Ariano, G.M., Perinotti, P., Valiron, B.: Quantum computations without definite causal structure. Phys. Rev. A **88**, 022318 (2013)
7. Araújo, M., et al.: Witnessing causal nonseparability. New J. Phys. **17**, 102001 (2015)
8. Predojević, A., Mitchell, M.: Engineering the Atom-Photon Interaction: Controlling Fundamental Processes with Photons, Atoms and Solids. Nano-Optics and Nanophotonics, 1st edn. Springer International Publishing (2016)
9. Chuu, C.-S., Harris, S.E.: Ultrabright backward-wave biphoton source. Phys. Rev. A **83**, 061803 (2011)
10. Eisaman, M.D., Fan, J., Migdall, A., Polyakov, S.V.: Invited review article: Single-photon sources and detectors. Rev. Sci. Instr. **82**, 071101 (2011)
11. Grangier, P., Roger, G., Aspect, A.: Experimental evidence for a photon anticorrelation effect on a beam splitter: a new light on single-photon interferences. Europhys. Lett. **1**, 173 (1986)
12. Brown, R.H., Twiss, R.Q.: Correlation between photons in two coherent beams of light. Nature **177**, 27–29 (1956)
13. Bocquillon, E., Couteau, C., Razavi, M., Laflamme, R., Weihs, G.: Coherence measures for heralded single-photon sources. Phys. Rev. A **79**, 035801 (2009)
14. Wolfgramm, F., de Icaza Astiz, Y.A., Beduini, F.A., Cerè, A., Mitchell, M.W.: Atom-resonant heralded single photons by interaction-free measurement. Phys. Rev. Lett. **106**, 053602 (2011)
15. Loudon, R.: The Quantum Theory of Light, 1st edn. Clarendon Press, Oxford (1973)
16. Luo, K.-H., et al.: Direct generation of genuine single-longitudinal-mode narrowband photon pairs. New J. Phys. **17**, 073039 (2015)
17. Hong, C.K., Ou, Z.Y., Mandel, L.: Measurement of subpicosecond time intervals between two photons by interference. Phys. Rev. Lett. **59**, 2044–2046 (1987)
18. Branczyk, A.: Non-classical states of light. Ph.D. thesis, University of Queensland (2010)
19. Xie, Z., et al.: Harnessing high-dimensional hyperentanglement through a biphoton frequency comb. Nat. Photonics **9**, 536–542 (2015)

Chapter 5
Conclusions

5.1 Summary

With this thesis, we accomplished an essential milestone towards an efficient light-matter interface, namely the creation of pure and indistinguishable single photon states from a bright quantum light source that are spectrally matched to be the optimal input of a gradient echo memory (GEM) [1–3], the most promising storage scheme to date.

We first summarise the theory necessary to design and characterise a narrowband single photon source based on SPDC. Understanding these elementary concepts is crucial to construct the device and avoid unwanted surprises in the building stage. The theoretical predictions for the single photon metrics, classical and quantum, give helpful insights into what to expect from the experiments and plan accordingly.

The thesis further presents the individual components of the source and preliminary characterisations, starting with the heart of the experiment: the optical parametric oscillator (OPO). It consists of a nonlinear crystal enclosed in an optical cavity and creates the desired output spectrum of narrowband modes at the rubidium D_1 transition line to be combined with GEM. The OPO is actively stabilised to the incoming pump light, which is created at exactly double the frequency of the atomic resonance. This is ensured by generating the pump light inside a resonant cavity, itself stabilised to the rubidium transition. Due to this cascading locking technique, we guarantee the frequency-degenerate central mode of the OPO to be on resonance with the atoms. We also give detailed insight into the experimental implementation of the frequency stabilisation steps and show the way the individual components are combined and how they work together.

Using cavity-enhanced SPDC to shape the emitted photons has a variety of advantages: spectrally, we demonstrate a linewidth of (429 ± 10) kHz by measuring the single photon intensity cross-correlation function $G_{s,i}^{(2)}(\tau)$. This is the narrowest SPDC-based single photon source to date, improving the former lowest achieved value [4] by a factor of four. The narrow linewidth combined with the central emission at the atomic transition of rubidium make this source an ideal candidate to be

© Springer Nature Switzerland AG 2018
M. Rambach, *Narrowband Single Photons for Light-Matter Interfaces*,
Springer Theses, https://doi.org/10.1007/978-3-319-97154-4_5

integrated with GEM. Additionally, triple resonant (signal - idler - pump) operation of the OPO allows 100% duty cycle, a feature where many narrowband sources fall short, and exceptionally high spectral brightness (SB) compared to other architectures [4–7]. We measure a background corrected SB of 1.13×10^6 photon pairs per second per mW pump power per MHz bandwidth. Taking into account the multi-mode operation of the source in this measurement, we infer (3900 ± 370) $\frac{\text{photon pairs / s}}{\text{mW} * \text{MHz}}$ in single-mode operation, surpassing the brightest source so far [8] by a factor of three. These classical features make the source well suited for quantum optics experiments demanding narrowband photons at high rates.

We furthermore demonstrate the quantum character of the created single photon pairs in two independent experiments. In the case of type II SPDC, the photons can easily be separated on a polarising beam splitter and send along two different paths. The detection of one photon (idler) can then herald the presence of a second photon (signal) in the other arm. In order to determine the multi-photon suppression of the single photon source, i.e. how close the signal photon is to a Fock state $|n = 1\rangle$, we measure the intensity auto-correlation function at zero time delay ($g_{s,s}^{(2)}(0)$). Our result of $g_{s,s}^{(2)}(0) = 0.032 \pm 0.003$ is 277 standard deviations below the classical limit of 1, and hence clearly shows the single photon nature of the source.

We determine the indistinguishability of the generated photons in a Hong-Ou-Mandel experiment, where two light particles interfere non-classically on a 50/50 beam splitter. The obtained interference visibility of (97 ± 3) % at zero time delay provides the high indistinguishability needed for efficient entanglement swapping schemes [9–13] in linear optical quantum computing and quantum networks. In multi-mode operation, the emitted photons have a unique comb-like spectral and temporal shape which results in revivals of the interference dips [14–16]. We demonstrate these revivals for a variety of different delays up to 105 m between signal and idler photons, far beyond anything reported in the literature so far and independently proving the long coherence length (narrow linewidth) of the photon pairs.

5.2 Outlook

There are a variety of short- and long-term goals we would like to achieve with the source, with some of them already on their way.

Firstly, we will complete the project on the HOM revivals. We have recently measured the dip revival at half an effective round-trip difference, but have yet to understand the results. Hence, the data is not included in this thesis, e.g. in Fig. 4.11g. This is ongoing work and should be completed within the upcoming month.

The main goal in the near future is certainly the demonstration of single-mode operation of the source. We are currently implementing a carefully designed triangular cavity, resonant with single photon and pump wavelengths in order to maintain the 100% duty cycle, into the existing setup. We ensure the overlapping of the individual resonances by adjusting the length of the cavity through a mirror mounted on a piezo-electric transducer and through miniature Peltier elements controlling the temperature of the incoupling mirror. Preliminary characterisation of this system has already proven its capabilities of up to ∼50% fibre-to-fibre transmission at 795 nm, but so far only for continuous laser light.

The primary long term aim of the project is the demonstration of an interface between our source and the gradient echo memory by our collaborators at the Australian National University in Canberra. Characterising the interaction of the narrowband single photons with hot or cold rubidium vapour in GEM by measuring e.g. storage efficiencies and times or recreating and comparing our single photon metrics after they have been stored and retrieved from the memory will give us great insight into the feasibility of the memory scheme. In order to achieve this demanding task, our source needs to be transported to Canberra. We have engineered the entire optical setup so that it fits on three small breadboards, all sized below 1×1 m, for high mobility. We have already demonstrated downtimes below two weeks for transits between different laboratories within our university and all necessary equipment like vibration absorbers and transport crates for long distance shipping have been purchased and delivered in anticipation of this transfer.

We believe that the techniques introduced in this thesis, especially the flip-trick, have the potential to become a standard in the field of narrowband single photon sources, in particular in devices where the clustering effect is not practical. The ability to create photons spectrally matched to atomic ensembles at a high rate and at any given time flings the door for efficient interfaces wide open. Additionally, the unfiltered multi-mode structure of the single photons could enable new methods of measuring long distances at sub-micrometer level accuracy.

While the main purpose of the source is its integration with GEM, the photons can also be utilised in other applications like quantum information processing and quantum foundations. We are planning to use the exceptional temporal and spatial length of the single photon wave packet in quantum foundation experiments. In order to perform quantum computations without definite causal structure, also known as the quantum switch [17, 18], it is essential for the photon to extend over the whole experimental setup to erase the path information that could be acquired otherwise. This is easily achieved with our photons, extending over more than 100 m. In this case, the qubit becomes entangled with the circuit structure, revealing unexplored aspects of quantum theory.

Another potential application is to send our photons through gas-filled hollow-core photonic crystal fibres [19, 20], developed by our colleagues at the University of Adelaide, in order to achieve high cross-phase modulations [21–23] or precise spectroscopy [24–26]. Particularly the cross-phase modulation is of high interest in the field of optical quantum computing, as engineering deterministic interactions between photons is challenging due to their weak interaction and needs to be mediated by a nonlinear medium [23, 27].

References

1. Hosseini, M., Sparkes, B.M., Campbell, G., Lam, P.K., Buchler, B.C.: High efficiency coherent optical memory with warm rubidium vapour. Nat. Commun. **2**, 174 (2011)
2. Hosseini, M., Campbell, G., Sparkes, B.M., Lam, P.K., Buchler, B.C.: Unconditional room-temperature quantum memory. Nat. Phys. **7**, 794–798 (2011)
3. Sparkes, B.M., Hosseini, M., Hétet, G., Lam, P.K., Buchler, B.C.: An ac stark gradient echo memory in cold atoms. Phys. Rev. A **82**, 043847 (2010)
4. Fekete, J., Rieländer, D., Cristiani, M., de Riedmatten, H.: Ultranarrow-band photon-pair source compatible with solid state quantum memories and telecommunication networks. Phys. Rev. Lett. **110**, 220502 (2013)
5. Rieländer, D., Lenhard, A., Mazzera, M., de Riedmatten, H.: Cavity enhanced telecom heralded single photons for spin-wave solid state quantum memories. New J. Phys. **18**, 123013 (2016)
6. Wolfgramm, F., de Icaza Astiz, Y.A., Beduini, F.A., Cerè, A., Mitchell, M.W.: Atom-resonant heralded single photons by interaction-free measurement. Phys. Rev. Lett. **106**, 053602 (2011)
7. Tian, L., Li, S., Yuan, H., Wang, H.: Generation of narrow-band polarization-entangled photon pairs at a rubidium d1 line. J. Phys. Soc. Jpn. **85**, 124403 (2016)
8. Chuu, C.-S., Yin, G.Y., Harris, S.E.: A miniature ultrabright source of temporally long, narrowband biphotons. Appl. Phys. Lett. **101**, 051108 (2012)
9. Duan, L.M., Lukin, M.D., Cirac, J.I., Zoller, P.: Long-distance quantum communication with atomic ensembles and linear optics. Nature **414**, 413–418 (2001)
10. Knill, E., Laflamme, R., Milburn, G.J.: A scheme for efficient quantum computation with linear optics. Nature **409**, 46–52 (2001)
11. Chou, C.-W., et al.: Functional quantum nodes for entanglement distribution over scalable quantum networks. Science **316**, 1316–1320 (2007)
12. Sangouard, N., Simon, C., de Riedmatten, H., Gisin, N.: Quantum repeaters based on atomic ensembles and linear optics. Rev. Mod. Phys. **83**, 33–80 (2011)
13. Munro, W.J., Stephens, A.M., Devitt, S.J., Harrison, K.A., Nemoto, K.: Quantum communication without the necessity of quantum memories. Nat. Photonics **6**, 777–781 (2012)
14. Shapiro, J.: Coincidence dips and revivals from a type-ii optical parametric amplifier. In: Nonlinear Optics: Materials, Fundamentals and Applications, FC7. Optical Society of America (2002)
15. Lu, Y.J., Campbell, R.L., Ou, Z.Y.: Mode-locked two-photon states. Phys. Rev. Lett. **91**, 163602 (2003)
16. Xie, Z., et al.: Harnessing high-dimensional hyperentanglement through a biphoton frequency comb. Nat. Photonics **9**, 536–542 (2015)
17. Araújo, M., et al.: Witnessing causal nonseparability. New J. Phys. **17**, 102001 (2015)
18. Chiribella, G., D'Ariano, G.M., Perinotti, P., Valiron, B.: Quantum computations without definite causal structure. Phys. Rev. A **88**, 022318 (2013)
19. Cregan, R.F., et al.: Single-mode photonic band gap guidance of light in air. Science **285**, 1537 (1999)
20. Benabid, F., Couny, F., Knight, J., Birks, T., Russell, P.: Compact, stable and efficient all-fibre gas cells using hollow-core photonic crystal fibres. Nature **434**, 488–491 (2005)
21. Bhagwat, A.R., Gaeta, A.L.: Nonlinear optics in hollow-core photonic bandgap fibers. Opt. Express **16**, 5035–5047 (2008)
22. Saha, K., Venkataraman, V., Londero, P., Gaeta, A.L.: Enhanced two-photon absorption in a hollow-core photonic-band-gap fiber. Phys. Rev. A **83**, 033833 (2011)
23. Perrella, C., et al.: High-efficiency cross-phase modulation in a gas-filled waveguide. Phys. Rev. A **88**, 013819 (2013)
24. Slepkov, A.D., Bhagwat, A.R., Venkataraman, V., Londero, P., Gaeta, A.L.: Spectroscopy of rb atoms in hollow-core fibers. Phys. Rev. A **81**, 053825 (2010)
25. Perrella, C., Light, P.S., Stace, T.M., Benabid, F., Luiten, A.N.: High-resolution optical spectroscopy in a hollow-core photonic crystal fiber. Phys. Rev. A **85**, 012518 (2012)

26. Perrella, C., et al.: High-resolution two-photon spectroscopy of rubidium within a confined geometry. Phys. Rev. A **87**, 013818 (2013)
27. Milburn, G.J.: Quantum optical fredkin gate. Phys. Rev. Lett. **62**, 2124–2127 (1989)

Appendix A
HWP Characterisation

The HWP is a core element of the single photon source. It is a half inch diameter, effective zero-order (cemented) Quartz retardation plate made for a π phase delay between orthogonal polarisations at 795 nm. The HWP sits in a high-precision rotation mount which allows fine rotation via an adjuster screw in the range of $\pm 7°$ around a set value. The characterisation of the HWP as a function of the angle of the rotation mount can be seen in Fig. A.1. Horizontally polarised light passes through the HWP and a polarising beam splitter before being measured on a power meter. Full transmission indicates that the HWP does not affect the polarisation while no transmission corresponds to a perfect flip from horizontally to vertically polarised light,

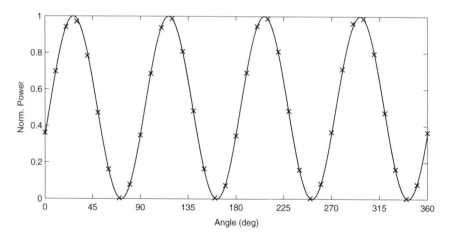

Fig. A.1 Characterisation of the HWP. The transmitted laser power through a polarising beam splitter behind the HWP is measured as a function of the HWP angle that can be read of the precision mount. Zero transmission at 72°, 162°, 252° and 342° corresponds to a perfect flip between two orthogonally linear polarisations. Error bars are smaller than the symbols representing the data

© Springer Nature Switzerland AG 2018
M. Rambach, *Narrowband Single Photons for Light-Matter Interfaces*,
Springer Theses, https://doi.org/10.1007/978-3-319-97154-4

the desired mode of operation. The $\sin^2$ fit to the data shows that this occurs at $72°$, $162°$, $252°$ and $342°$. Out of the four possible options the angle was roughly adjusted to $72°$ and then fine tuned to minimise the odd number of round-trip differences for perfect alignment, as seen in Fig. 4.2 of Sect. 4.1.1.

Appendix B
Photodetector Circuit Diagram

See Fig. B.1 and Table B.1.

© Springer Nature Switzerland AG 2018
M. Rambach, *Narrowband Single Photons for Light-Matter Interfaces*,
Springer Theses, https://doi.org/10.1007/978-3-319-97154-4

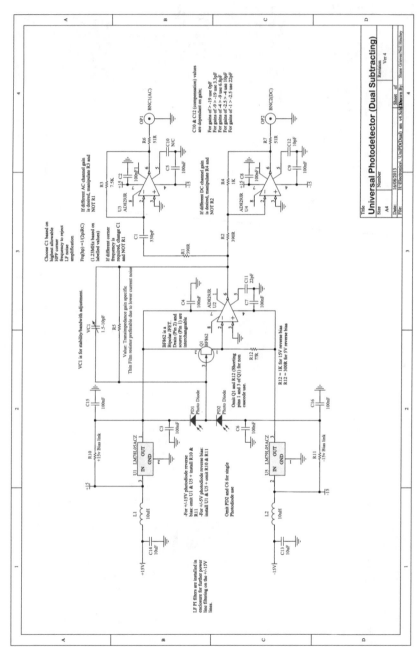

Fig. B.1 Circuit diagram of the universal photodetector, used in various locations throughout the setup. Figure from [1]

Table B.1 List of parts necessary to build the universal photodetector. pF, picofarad; nF, nanofarad, μF, microfarad; μH, microhenry; Ω, ohm; kΩ, kiloohm; V, volt; C, capacitor; L, inductor; R, resistor; U, integrated circuit; VC, variable capacitor; PD, photodiode

Comment	Designator
OP1	BNC2
OP2	BNC2
330 pF	C1
100 nF	C2
100 nF	C3
100 nF	C4
100 nF	C5
100 nF	C6
100 nF	C7
100 nF	C8
100 nF	C9
Variable between 0 and 22 pF	C10
22 pF	C11
Variable between 0 and 22 pF	C12
10 μF	C13
10 μF	C14
100 nF	C15
100 nF	C16
10 μH	L1
10 μH	L2
390 Ω	R1
390 Ω	R2
7.5 kΩ	R3
1 kΩ	R4
2 kΩ [Thin Film] (dependent on optical power)	R5
51 Ω	R6
51 Ω	R7
1 kΩ for 15 V or 300 Ω for 5 V	R12
0 Ω for 15 V bias	(+) 15 V link
0 Ω for 15 V bias	(−) 15 V link
LM78L05ACZ (for 5 V)	U1
AD829J	U2
AD829J	U3
AD829J	U4
LM79L05ACZ (for 5 V)	U5
1.5–10 pF	VC1
Hamamatsu S3883, S5973-01/02	PD1
10 pF	C10 and C12 alt
6.8 pF	C10 and C12 alt
3.3 pF	C10 and C12 alt

Reference

1. Stefszky, M., Gmeiner, J.: ANU Photodetector V4. Technical Report, Australian National University, Canberra, Australia (2014)

Curriculum Vitae

Markus Rambach is an experimental quantum physicist from Lustenau, Austria. He specialises in the hybridisation of quantum technologies for the implementation of quantum networks. He is particularly interested in the engineering of interfaces between photons (flying qubits) and atoms or ions (stationary qubits), an essential step towards the next generation of the internet. Using specifically developed quantum light sources, his research is driven by the possibilities of quantum networks to facilitate secure communication in the future.

He completed his Bachelor and Master of Science degrees in quantum physics at the University of Innsbruck, Austria, where he learned the crucial elements of the quantum optics toolbox and was involved in building a universal digital quantum simulator. He received two scholarships from the University of Queensland (UQ), Australia, to perform his postgraduate research and was awarded his PhD in 2017.

Outside the laboratory, he enjoys teaching and communicating his research to the scientific community. As part of the UQ student chapter board of the Optical Society of America, he co-organised multiple student conferences and social events. He is also involved in an academic initiative aiming to improve educational outcomes for Aboriginal and/or Torres Strait Islander students.

© Springer Nature Switzerland AG 2018
M. Rambach, *Narrowband Single Photons for Light-Matter Interfaces*,
Springer Theses, https://doi.org/10.1007/978-3-319-97154-4

Employment History and Education

Research Associate Jun 2018–present
Heriot-Watt University, UK; Focusing on hybrid quantum technologies.
Postdoctoral Research Fellow May 2017–Dez 2017
The University of Queensland, Australia and The Australian National University, Australia; Focusing on the hybridisation of quantum systems.
Doctor of Philosophy Mar 2013–Apr 2017
The University of Queensland, Australia; Thesis title: *Narrowband Single Photons for Light-Matter Interfaces*.
Treasurer of the UQ OSA Student Chapter Jan 2014–Dez 2016
Organised three student conferences and various social events for students.
Master of Science Oct 2009–Nov 2011
University of Innsbruck, Austria; Thesis title: *Laserkühlung von Ionenkristallen* (*Laser Cooling of Ion Crystals*).
Bachelor of Science Oct 2005–Sep 2009
University of Innsbruck, Austria; Specialising in experimental physics.

Professional Awards

Best Poster Award in Physics 2015
The University of Queensland, Australia.
International Postgraduate Research Scholarship 2013–2016
The University of Queensland, Australia.
Australian Postgraduate Award 2013–2016
The University of Queensland, Australia.

List of Publications

1. **Hectometer revivals of quantum interference**,
   M. Rambach, W. Y. S. Lau, S. Laibacher, V. Tamma, A. G. White, T. J. Weinhold
   arXiv preprint https://arxiv.org/abs/1806.00013 (2018).
2. **Sub-megahertz linewidth single photon source**,
   M. Rambach, A. Nikolova, T. J. Weinhold, A. G. White *APL Photonics* **1**, 096101 (2016).
3. **Universal Digital Quantum Simulation with Trapped Ions**
   B. P. Lanyon, C. Hempel, D. Nigg, M. Müller, R. Gerritsma, F. Zähringer, P. Schindler, J. T. Barreiro, M. Rambach, G. Kirchmair, M. Hennrich, P. Zoller, R. Blatt, C. F. Roos *Science* **334**, 57–61 (2011).
   ISI Highly cited paper, Top 1% most cited papers in physics (ISI).

Printed in the United States
By Bookmasters